The In-Home
VCR
Mechanical Repair
& Cleaning Guide

The In-Home
VCR
Mechanical Repair
& Cleaning Guide

By Curt Reeder

PROMPT.
PUBLICATIONS

An Imprint of
Howard W. Sams & Company
Indianapolis, Indiana

FIRST EDITION, 1996

PROMPT® Publications is an imprint of Howard W. Sams & Company, a Bell Atlantic Company, 2647 Waterfront Parkway, E. Dr., Suite 300, Indianapolis, IN 46214-2041.

International Standard Book Number: 0-7906-1076-0

Library of Congress Catalog Card Number: 96-67585

Acquisitions Editor: Candace M. Drake
Editor: Natalie F. Houck
Assistant Editors: Karen Mittelstadt, Pat Brady
Illustrator: Curt Reeder
Additional Illustrations By: **PRB Line®**
PO Box 28
Whitewater, WI 53190-0028
1-800-558-9572
Typesetter: Natalie Houck
Cover Design: Christy Pierce
Cover Art Supplied By: **Sencore**
3200 Sencore Dr.
Sioux Falls, SD 57107-0000

Printed in the United States of America

9 8 7 6 5 4 3 2 1

TABLE OF CONTENTS

I'd like to extend a special thank you to the following people,
who helped to bring this book into being:

My parents, Wanda and Ernest Reeder
The staff at PROMPT Publications and Howard W. Sams & Company
John Dwinell, PRB Corporation
Paul Szots, Video Care and Repair of Salem, NH
William Zappala, Andover Video
John Daly, Daly Video
Geraldine Sasso, Stop -N- Go Video
John Leonard, Photo USA

Richard, Jennifer, Cailen Graber, Bruce, Ruth, Scott, Doug Baker, Graham, Louisa,
Ardith, Eve LaRochell, Vanessa Gilbert, Debbie Callina, Jane Gossard,
Marty Roach, and all customers of "In-Home VCR."

Preface

Preface

The video cassette recorder, or VCR, has become a household fixture. It is listed with televisions and stereos in nearly every Sunday morning department store sales flyer in newspapers across the country. Rarely has a generation witnessed such an explosion of high technology made affordable enough to sit in nearly every home in the United States.

If ever there was a "television generation," this is certainly the "VCR generation." To date, there are nearly 78 million VCRs on the market. Ask a group of people if they own a VCR, and chances are that one person out of four will have one. With the exception of viewing television, viewing rental movies via VCR has become a national pastime.

VCRs have made their niche in the nine-to-five world as well. There are many uses for VCRs in today's work force. Some uses include reviewing real estate properties, bank security systems, business training films, educational classes, weather forecasting, dating services, and of course, use by Hollywood studios.

Twenty years ago, if you told me I would someday be able to view any program recorded from television and review it over and over again at my leisure, I would have laughed first and asked for proof later. Now the proof sits on or around our television sets.

This VCR repair and cleaning manual is geared toward *you*, the average home VHS VCR user who owns one or more VCRs, and would like years of tested service technician information to assist you in maintaining your equipment on a regular basis. Like anything that is used in your home (vacuum cleaners, lawn mowers, etc.), a VCR requires minimal service to keep functioning well. The tools used to maintain this appliance can be found in a kitchen junk drawer or purchased at a local hardware store. It cannot be stressed enough that a technical or electrical engineering degree is NOT required for average home users to begin maintaining VCRs on a regular basis.

When a VCR does develop a symptom, it is most often due to one or two minor faulty parts or simple dirt buildup from normal use. VCR problems can be classified into three categories: electrical, mechanical, and electro-mechanical. This text concentrates on the most common VCR faulty symptoms in any one of these categories, and illustrates in non-technical terms the proper method to correct these problems.

Informing the VCR owner of the many variables to look for and how to eliminate particular problems makes this book a useful tool. With only a few small hand tools such as tweezers, cleaning fluid, a power screwdriver and cotton swabs, today's VCR owner is well on the way to learning the many tricks and secrets of VCR maintenance.

This book is also geared toward the entrepreneur who may consider starting a VCR service business of his/her own. The vast information contained herein gives the novice VCR service person a firm foundation on which to create a personal niche in this unique service business.

In 1990, I started *In-Home VCR Repair and Cleaning*, servicing VCRs by appointment in the homes of customers. Since starting this small service business, I have compiled many notes and descriptive remedies that illustrate, in easy-to-understand terms, how the owner of one or more VCRs can service their own units and possibly start a small business.

I've tried my best to make this the guide I wish I'd had when I first started servicing VCRs. Documented customer service calls of the most common symptoms have brought this book into being. The owner of this book will quickly identify the numerous problems which affect all VCRs on the market today. Beyond this, they will be grateful to have found a solution to their problem within the same text.

Although there are many VCR repair books on the market today, none seem to get to the heart of faulty VCR symptoms as quickly and easily as this book. Competitive books on the market today tend to avoid (for reasons I have yet to comprehend) real solutions to common problems which affect every VCR. Either the contents of these publications are too advanced and geared exclusively toward the professional electrical engineer, or are too simple, seemingly written for humor at the expense of the book purchaser. In any case, the reader is often left more confused than when they began reading the book.

I have been in the VCR service business for several years, and I have correctly repaired literally hundreds of VCRs, by appointment only, with minimal rework or negative customer response by using the methods outlined within this text. It is safe to say that this book is tried-and-true to its readers. The most common and frequent VCR malfunctions that customers have experienced have brought this book into being.

This book presents repairing and cleaning VCRs as a step-by-step process. Each chapter of this book lends more information to the next chapter, at which point the reader can actively search for the cause behind the problem with their VCR and understand how to correct it.

I hope you enjoy the presentation of this information and find yourself referring to the text again and again in the future to correct many VCR problems.

Introduction
SIX RULES OF THE GAME

Introduction
SIX RULES OF THE GAME

Before you actually begin to disassemble your working VCR in an attempt to correct its original symptom, there are several rules which should be taken into consideration in order to make your first disassembly, repair, cleaning and reassembly process more enjoyable and glitch free.

Rule #1

Never attempt to work on your VCR or any other VCR with its power supply cord plugged into a wall outlet. There is a warning label on the rear panel of every VCR manufactured which states, "TO AVOID AN ELECTRICAL SHOCK, DO NOT REMOVE THE COVER." This warning should be taken very seriously. By removing the power supply cord from any power supply outlet, you should remove any possibility of a hazardous shock.

Rule #2

WARNINGS have been provided throughout this text at key intervals which are intended to warn you against a potential hazard or to prevent you from causing another, more costly fault in the VCR you are working on. The WARNINGS should be taken very seriously and should be read several times before you continue with the actual disassembly or cleaning procedure.

Rule #3

It is highly recommended that you purchase an unwanted "junker" VCR to practice on before attempting a cleaning or repair procedure on a working VCR that may be exhibiting a fault. A junker VCR is a great learning tool and will allow you to make an unlimited number of mistakes without the pressure of a friend, family member or customer anxiously awaiting word about the condition of their VCR.

Rule #4

If a repair or cleaning procedure seems beyond your ability to correct, then don't do it. It may be best left to a friendly VCR service center. Be certain to make a note of what the VCR technician had to do in order to correct the original symptom. Use this text and try to decide where the problem occurred. By making a note of the corrective procedure, you may be able to correct the problem yourself if you happen to see the same symptom again in the future. Also, by not attempting to correct the problem yourself, you avoid an otherwise inflated repair bill required to correct any misguided "tweaking".

Rule #5

Always make a point to use manufacturer's original replacement parts when performing a repair procedure. By using the manufacturer's parts, you will avoid second-guessing your original repair work and reduce the chance of a negative customer response. As you will see, although VCR parts tend to look similar in size and shape, it is not necessarily true that they are interchangeable.

Rule #6

As you will see in the following chapters, the proper disassembly steps outlined in this text should be applied to every VCR you happen to work on. The proper order is:

1. Remove the VCR's lid.
2. Remove the VCR's lower pan.
3. Remove the VCR's front panel.
4. Remove the VCR's head shield.
5. Remove the VCR's carriage assembly.

Keep in mind, as you progress through this method, that each disassembly step is considered a judgment call. That is to say, removing the VCR's lid may be all that is required in order for you to remove something from within the VCR's mechanical format in order to remedy an original faulty symptom. Soon you will be able to judge for yourself as to how far the VCR will need to be disassembled in order to access the faulty or dirty part that is creating the original symptom.

The secret to VCR repair and cleaning is quite simple:

1. Go slow at each disassembly stage.
2. Understand the function of each part in the typical VHS VCR mechanical format.
3. Obtain the ability to recognize the symptoms of a particular part when it is dirty or fails.
4. Use the proper disassembly steps outlined in this text to clean, adjust or replace the part which is exhibiting the symptom.

Also, as a reminder, there are two terms that need to be clarified which are referred to within this text. The two terms are:

1. VHS VCR format.
2. Symptom similarity.

Let's clarify the term "VHS VCR format" first. In the early 1980s, there were two distinct formats of VCR available to the purchasing public. They were VHS and Beta. At this time, the Beta format has all but become extinct while the VHS format has become the norm. Since this text is based on the VHS VCR, the term "format" refers to the VHS VCR format and the general inner mechanical workings of the VCR, rather than the difference between Beta and VHS. For example, the mechanical format of Fisher brand VCRs is quite different from the mechanical format of RCA brand VCRs, though they are both VHS standard.

This is an important aspect to take into consideration when considering the original symptom in an ailing VCR. An experienced VCR technician will often ask, "What brand is your VCR?" Since each VHS VCR tends to have its share of likely sources for a particular fault to occur, this question is asked first in the troubleshooting process to more quickly locate the origin of the symptom affecting the VCR in question.

The second term to consider is "symptom similarity". Symptom similarity is when two or more VCR parts exhibit the same symptom. For example, the symptom for failed video heads looks very similar to the symptom for a guidepost that may be obstructed in its groove. However, the difference in price to correct a guidepost is quite minimal compared to replacing the video heads.

This, too, is an important aspect to keep in mind when troubleshooting a VCR. Although it may seem as though there are over a thousand different variables to take into consideration, once you understand the many different parts of a VCR and how they are intended to perform, often the origin of a symptom can be correctly identified after considering only three or four different parts within the typical VCR.

Keep the above six rules in mind, and let's begin by first outlining the common parts of the trade which you will most likely encounter as you begin to repair VCRs for friends, family, and a handsome profit.

Chapter One
GETTING STARTED

Chapter One
GETTING STARTED

It seems that the smallest part of a VCR tends to do the most work. Call it planned obsolescence, call it bad karma; at any rate, VCRs tend to fail, and they often don't provide adequate warning when they do. When a VCR does begin to exhibit a symptom, chances are a part most likely has failed or has accumulated dirt through many hours of normal use.

It is not essential that you understand the purpose of these parts or their names for that matter. Simply note the different size and shapes of the common VCR parts. An explanation of each part will be given in detail as that particular part comes into focus during the disassembly, cleaning or repair process of the typical VCR.

1.1 Parts of the Trade

As can be seen in *Figure 1-1*, VCR parts tend to include idler drive assembly tires, pinch rollers, tape tension bands, various side drive belts, and an assortment of nylon gears. If you think you are going to be servicing VCRs on a fairly regular basis, it's a good idea to have several replacements for these parts which are compatible with different models of VCRs.

Figure 1-1. A collection of common VCR parts.

At this time, you should begin to establish an array of both large and small hand tools that you will use often in order to put your VCR in a serviceable position. These hand tools can be purchased at your local department store for less than $50.

1.2 Large Hand Tools

Of the large hand tools (*Figure 1-2*), your collection should include a set of screwdrivers of various sizes, a pair of wire cutters, needlenose pliers, a soldering iron, and a power screwdriver.

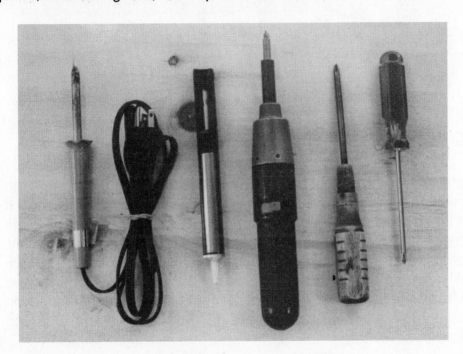

Figure 1-2. Large hand tools.

1.3 Small Hand Tools

The category of small hand tools should include a nine volt battery, plastic probes, tweezers, a dental mirror, small Phillips and flathead screwdrivers, bonding cement, a spring release hook, a permanent marker, and an infrared detector (*Figure 1-3*).

1.4 Cleaning Materials

Aside from the various large and small hand tools, cleaning materials are important to keep on hand. Be certain to purchase cotton-tipped cleaning swabs, cleaning solution (a mild fingernail polish remover works well), video head cleaning sticks, and a leather chamois (*Figure 1-4*).

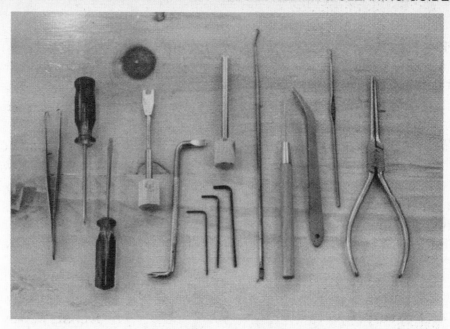

Figure 1-3. Small hand tools.

1.5 Your Work Area

The first step toward performing a repair or cleaning procedure is to set aside an area in your household as a designated work space. It cannot be stressed enough that a proper work area will often be the deciding factor in whether or not your repair attempt goes smoothly. There are several work space criteria which should be met to ensure that further damage does not result from disassembling and reassembling your VCR. If this list of requirements is adhered to, you should have an ideal work space for your repair and cleaning procedures.

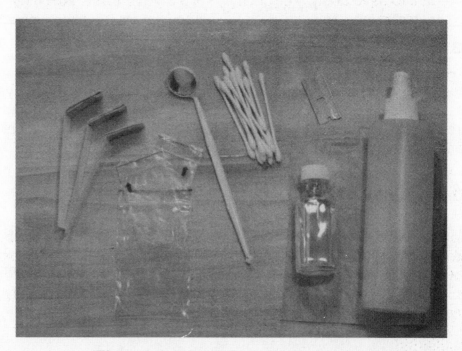

Figure 1-4. Commonly used cleaning materials.

1.5.A List of Work Space Requirements:

1. When choosing a proper work area, be certain your work space is not in the way of common foot traffic. This includes, but is not limited to, wet pets, curious children, rambunctious house guests, etc.

2. The surface that you choose to use as a work table should be flat and nonconductive. A plywood or Formica table top works very well for this application. Your VCR should always sit flush with the table top.

3. The third aspect to take into consideration is any potential overhead hazards. These hazards may go unnoticed at first, until tragedy strikes. Avoid setting up your work area where there are overhead water pipes, open windows, or stacks of teetering books.

4. It is very important that your work space has adequate power supply outlets in order to provide electricity for such things as your soldering iron, power screwdriver recharger, desk lamp, television, and the VCR that you are currently working on. It is recommended that you purchase a six-outlet power surge extension cord. By utilizing a multi-outlet power surge extension cord, the items that you use most often (such as the television, desk lamp, and power screwdriver recharger) can remain plugged in when they are not in use. This type of outlet will also reduce the chance of accidentally plugging in a VCR that has been disassembled, resulting in a short circuit, when you really intended to plug in the power cord to a desk lamp.

5. Aside from the previous four work space criteria, it is important to have adequate lighting. Keep in mind that a VCR may automatically power off if the overhead light is too direct once its lid has been removed. A desk or wall lamp should suffice for your first VCR repair or cleaning procedure.

6. As a reminder, if you were to seriously consider opening a VCR repair service, it is very important to organize the repaired VCRs on proper shelving with adequate spacing between each shelf. Never stack VCRs on the floor or stockpile them in a corner. This is a sure sign of a "Sloppy Joe".

1.6 The Television as a Test Tool

Once you have decided on a good location to perform your VCR repair and cleaning procedures, it is a good idea to use a secondhand color television as a test tool. A color monitor is essential for testing a VCR's recording capability as well as its color playback clarity in each of its various playback speeds (SP, LP and SLP).

Refer to Chapter Ten for making the proper television antenna-to-VCR hookup in order to receive broadcast programming through the VCR you are currently working on.

After the television antenna has been connected correctly, you should be left with two cables running to the back of the VCR, a cable for "Antenna In" and a cable for "Out to TV." These are the only two cables that have to be connected for each VCR you hook up to your color television, in order to test the VCR's record and playback picture quality (*Figure 1-5*).

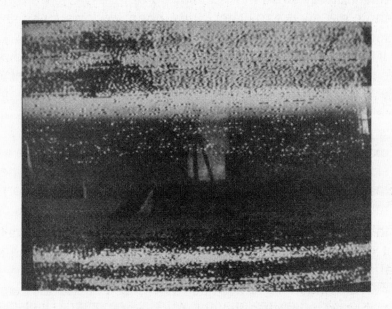

Figure 1-5. You can test a VCR by examining the playback picture on the TV screen.

1.7 Soldering Irons

Of the larger hand tools, the soldering iron will be one hand tool that is used extensively in order to perform some of the more advanced disassembly and replacement procedures outlined in future chapters. Learning how to solder electrical components to and from the VCR's electrical format is very easy to learn, and requires only fifteen minutes of practice to master.

The materials required are a 15 watt soldering iron, solder, and a desoldering braid (*Figure 1-6*). It is very important not to use a soldering iron that is greater than 15 watts. A soldering iron of greater wattage may severely overheat the new electrical component that you are attempting to install, rendering it useless. Often, a reliable soldering iron can be purchased for less than $10.

Figure 1-6. *A soldering iron, solder, and desoldering braid.*

1.8 Solder Off

Begin this lesson by first plugging your soldering iron into a wall outlet since it requires several minutes for the tip of the soldering iron to preheat to its proper temperature. It is recommended that a junker VCR or an unwanted appliance such as a radio that contains electrical circuits be used in order for you to experiment. This will allow you to make an unlimited number of mistakes without the anxiety of potentially destroying the electrical components in a perfectly good VCR.

After several minutes, test to see whether the tip of the soldering iron has heated to its proper melting point by touching a small length of solder to the tip. The solder should melt quickly once it comes in contact with the soldering iron, indicating that the iron is ready.

Once the tip of the soldering iron has been preheated to its correct temperature, simultaneously touch the side of the lead wire to the electrical component you are attempting to remove, with the tip of the soldering iron over the desoldering braid. The solder should quickly melt, and will be absorbed into the braid until all of the solder has been removed and the faulty component can be extracted from the VCR you are currently working on (*Figure 1-7*).

WARNING: *Be certain not to touch the tip of the soldering iron once the soldering iron has been plugged into a wall outlet. The tip of the soldering iron becomes VERY hot and should be kept out of the reach of children. NEVER touch the tip of the soldering iron with your finger as a test to see whether or not it is ready to melt solder; use a length of solder instead. Once the soldering iron has reached its proper melting point, the test piece of solder should melt quickly once it touches the tip of the soldering iron.*

Figure 1-7. *Desoldering a component from a PC board.*

1.9 Solder On

After desoldering and removing the faulty electrical component, try installing it back into the junker VCR you are using.

In order to solder components into the VCR, simply insert the new part (in this case, the part you just removed since this is just a practice lesson) back into the PC board. Simultaneously, touch the lead wire of the new component with both the tip of the soldering iron and a small length of solder. Once the solder makes contact with the heated tip of the soldering iron, the solder will quickly melt into place around the base of the lead wire of the electrical component (*Figure 1-8*). Remove both the soldering iron and solder. The solder should harden into place around the lead wire.

These are rather simple explanations of soldering and desoldering. Keep in mind that there are certain things to consider while soldering, such as what kind of solder to use, what a good solder joint looks like, care of the circuit board and adjacent paths, etc. For more information, get advice from your local parts dealer, or consult the PROMPT Publications book, *Surface-Mount Technology for PC Boards*.

Figure 1-8. Soldering a component to a PC board.

1.10 Tricks of the Trade

Here are some common "tricks of the trade" that you can use while servicing your VCR. Be sure to consult them often as you encounter problems while working on your VCR.

1.10.A Screw on a Stick

When reassembling your VCR, many times you will have to return a small mounting screw to a location within the VCR that is very difficult to access. A good example of this might be the base mounting screw which secures the carriage assembly to the VCR's metal chassis.

It is VERY frustrating to have the mounting screw tumble into the mechanical workings of your VCR after the VCR has been correctly repaired and the reassembly process is 99% finished. The result is often another forty minutes of disassembly in order to remove the lost mounting screw. There is an easy way to get the screw to its proper location without the risk of losing it in the chassis.

First, begin by trimming a six-inch wooden cleaning swab so that it can be inserted into the head slot of the screw. Next, simply fit the mounting screw securely on cut end of the wood swab. It is very important to be certain the mounting screw fits securely on stick before placing it over the open body of the VCR. Once you are certain it will not fall off, lower the mounting screw into its proper location within the VCR and begin to rotate the swab until the mounting screw begins to thread into its hole. Once the mounting screw is securely in its hole, remove the stick and tighten the screw with a screw driver.

1.10.B Drop of Oil on a Stick

At some point during a typical VCR repair and cleaning procedure, it may be necessary to apply a small amount of sewing machine oil to a part for lubrication in order to correct a loud squeak or to reduce mechanical friction between two tape path parts.

Because the parts that make up the typical VCR are often very compact, deliberately tipping a can of sewing machine oil over an open VCR in hopes of hitting the correct part would only shower the entire VCR with unwanted and potentially hazardous oil. This trick will assist in directing the oil to the part or gear for which it is intended.

Snip off the cotton pad of a six-inch wooden cleaning swab. Position the cut end of the wood swab on the gear or part which you intend to oil. Be certain to hold the wood at a slight angle.

Next, slowly apply one drop of sewing machine oil to the top of the swab. The droplets of oil will eventually accumulate and run the length of the swab into the awaiting gear or part location. Once the intended area has been sufficiently lubricated, discard the swab to ensure that other parts within the VCR are not contaminated by oil.

1.10.C Video Head Tape-Speed (SP, LP and SLP) Test Cassette

A tape speed test cassette is commonly found in VCR service centers. It allows the VCR technician to test the quality of the video heads in each of the various playback speeds: SP, LP and SLP. It is easy for the novice VCR repair and cleaning person to make their own head speed test cassette.

It is very important to use a VCR that is known to have good video heads and can record well in each playback speed. Also, it is a good idea to use a new video cassette that you know for certain has not been damaged by a VCR, in order to avoid transferring dirt from your test cassette to the video heads of another VCR.

Begin by recording several minutes of standard broadcast television programming in each of the VCR's recording speeds : SP, LP and SLP. It is very important that the television reception received through the recording VCR is clear and free of any "snow" or static. Record about three minutes worth of tape at each speed. Once you have recorded enough footage, *congratulations!* You have successfully made your own video head tape-speed test cassette. It will now be easy to test the playback picture quality of any VCR that you suspect has dirty or faulty video heads. To do this, simply insert your test cassette into the VCR, rewind, then press the Play command.

As the test cassette plays, notice that the tape speed will change from SP to LP then finally to SLP. Try to detect when the speed changes from one to another by observing the playback picture quality at each particular playback speed. Often, the playback speed that exhibits a field of "snow", "bullets," or "meteors" is the defective speed. This is an indication that the video heads may need to be cleaned.

If the snow, bullets or meteors are still present in any of the playback speeds after the video heads have been thoroughly cleaned, chances are that the video heads are faulty and need to be replaced.

1.10.D Dummy Jig Test Cassette

During the course of troubleshooting VCRs, it will be convenient to use a jig test cassette in order to observe the mechanical parts of the VCR while it is in Play, Record or Rewind mode. A jig test cassette is simply an unwanted video cassette (such as a retail video head cleaner) that has had its inner spools of videotape removed; what remains is only the plastic video cassette shell. Jig test cassettes are used extensively in VCR service centers to "fool" the VCR into thinking that an actual videotape has been entered. When this occurs, the front panel commands will respond, thus allowing the user to observe an idler drive tire that is not rewinding, a tape tension band that has come loose, or a pinch roller that does not press up against the capstan post; symptoms within the VCR.

1.10.E Sandwich Bags as Organizers

When disassembling your VCR, it is always a good idea to use labeled sandwich or storage bags to organize the many small parts and mounting screws generated from the disassembly and cleaning procedure (*Figure 1-9*).

Some people suggest the use of egg cartons, spray paint can tops and ice-cube trays as organizers. These are not recommended for one important reason: if your work area is disrupted and the various trays are sent crashing to the floor, chances are you may never be able to locate a missing tension spring, an integral locking washer, or a mounting screw. Believe me, sandwich bags are the way to go.

Figure 1-9. Sandwich and storage bags make good part organizers.

It is a good idea to label each sandwich bag based on the location within the VCR from which the loose parts came. For example, lower pan mounting screws should be set aside in a sandwich bag labeled "Lower Pan," etc. This method saves hours of guesswork and accelerates the learning process. Once you understand one VCR, the rest are typically the same.

1.10.F The Nine-Volt Jolt

In the course of troubleshooting VCRs, it is common to run across a VCR with a video cassette still within the carriage assembly. In most cases, the stuck video cassette will need to be removed before you can correct the problem and prevent this "stuck tape" symptom from occurring again.

Although the ailing VCR may not respond to the Eject command, it is often possible to use a nine-volt battery on the carriage assembly motor to expel the stuck video cassette.

To do this, simply touch the positive (red) and negative (black) contacts of the nine-volt battery to the appropriate lead wires on the carriage assembly loading motor. Unless the loading motor is faulty or the carriage assembly has a broken gear, the moment the battery comes in contact with the motor, it should jump to life and eject the video cassette. Only after the video cassette has been successfully ejected can the appropriate disassembly steps be taken.

1.10.G Useful Symbols as Guidelines

It is always a good idea to mark your course within the VCR with certain markings when you are attempting to disassemble it. By placing a small "T" to denote the topside of a gear, a hash mark (/) next to a particular mounting screw, the letters "Rd" to denote the location of a red mounting screw, or the letters "grd" to denote a ground wire, you save yourself a lot of guesswork when it is time to reassemble your VCR.

An orientation arrow drawn on the lower pan pointing to the back of the VCR should also be used to assist in denoting which end of the lower pan is either up or down. This saves a lot of time when the lower pan is temporarily in place to test the performance of the VCR.

By using "useful symbols," you train yourself in the placement of certain parts. Through practice, you will soon find yourself breezing through more and more difficult disassembly procedures which would have been confusing had you not initially used this personalized marking system.

1.10.H The Soldering Iron Coil

At times, during a part replacement procedure, it may be necessary to desolder a component from the VCR that is otherwise difficult to access. This trick of the trade assists in accessing the part while at the same time avoiding damage to the board on which the component is mounted.

Figure 1-10. *Using a soldering iron coil to increase the reach and precision of a soldering iron.*

To make a soldering iron coil, simply strip the insulation from a five-inch piece of heavy gauge wire. Before plugging in the soldering iron, wrap the gauge wire around the base and tip of the cool iron so that an inch of wire extends from the tip of the soldering iron. Next, plug the soldering iron into a wall outlet and let it heat to its proper temperature. The extended piece of wire is now the new soldering iron tip (*Figure 1-10*). You should find that the extended tip of the gauge wire can access a smaller area to desolder, thus avoiding any damage due to burning or overheating the VCR's PC board.

1.10.I X-Nut Adjustment Tool

When performing an audio control erase head adjustment, you will need to use a specific adjustment tool designed to fit into the top of the X-nut located at the base of the ACE head, while the VCR is playing a videotape. Aside from the hazards of reaching into the VCR while it is in use, there is one significant draw back when using the ACE head adjustment tool: it is often too short to be effective.

To remedy this problem, consider using a large flathead screw driver that has a notch cut to the same dimension as the manufacturer's X-nut adjustment tool. The added length of the screw driver will provide more leverage for turning the X-nut, and will reduce the chances of a tool being dropped into the rotating VCR video heads or passing videotape.

Figure 1-11. Shag rug dust remover.

1.10.J The Lower Pan Shag Rug Dust Remover

In the early '70's three things were popular : shag rugs, bean bag chairs, and disco. If you are the type of person that can't part with nostalgia, chances are you have a square piece of pea-green shag rug lying around. This could be a handy tool for removing dust from your VCR.

The lower pan of any VCR is notorious for collecting dust and dirt. Often, when the VCR's lower pan is removed, it is for the first time in the VCRs life. You can make a dust remover by simply securing a square foot of carpeting to a piece of plywood, mounted at a slight angle. To remove accumulated dust, spray a small amount of glass cleaner onto the piece of carpeting and run it across the lower pan (*Figure 1-11*). After many cleanings, simply discard the carpeting and begin with a new piece. This trick of the trade saves on both paper towels and cleaning solution.

1.10.K Nail Polish as "Glip"

When servicing a VCR, it may be necessary to perform an adjustment procedure in order to rectify a symptom. This adjustment procedure may require turning a screw to a specific location or setting the tape tension band to the proper tension. It is a good idea to use common nail polish as "glip" or adhesive to secure mounting screws, X-nuts and tension bands once they have been adjusted to their proper setting, so that they do not slip again after the VCR is reassembled.

Chapter Two
THE VCR'S SERVICEABLE POSITION

Chapter Two
THE VCR'S
SERVICEABLE POSITION

At this point you should have an adequate work space which has both proper lighting as well as ample power supply outlets in order for you to begin servicing VCRs. You are now ready to begin the initial steps toward disassembling, diagnosing, cleaning and repairing your junker VCR, and any other VCR on the market today.

Manufacturers of VCRs have taken into consideration that all VCRs require servicing at some point in the future. With more than 78 million VCRs in use today, a reliable VCR service technician could not be more in demand. In order to replace, clean or adjust the various mechanical parts in the typical VCR, VCRs are designed to go into what is called their "serviceable position". This means that the VCR's lid, lower pan, front panel, head shield, and carriage assembly are removable in order to access the dirty or faulty part in question.

This chapter is designed to walk the novice VCR owner through the proper steps toward putting a VCR into its serviceable position. Because there are so many different VCRs on the market, it would be impossible to cover each model in detail. However, once you understand the correct disassembly steps and understand what to look for at each particular stage, it is possible to apply this method to any VCR.

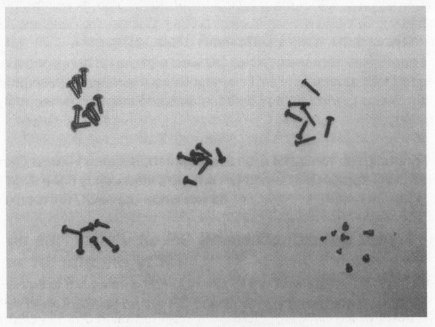

Figure 2-1. *Various types of VCR mounting screws.*

Before removing the first of many mounting screws from either your junker VCR or a faulty VCR, it is a good idea to read up on the different types of mounting screws used in VCRs on the market. This will give you a good idea of what to expect.

Typically, there are five different categories of mounting screws which you will see in the course of disassembling any VCR. It is very important to return mounting screws to their proper location after a repair or cleaning procedure to ensure another symptom does not occur.

2.1 Mounting Screw Categories

The five different mounting screws are lid mounting screws, lower pan mounting screws, red mounting screws, inner mounting screws, and very small mounting screws (*Figure 2-1*).

Lid mounting screws are often the largest screws removed from the VCR during a standard repair or cleaning procedure. Although chassis mounting screws are similar in size and shape, it is not likely that these particular mounting screws will be removed during a standard repair or cleaning procedure.

Lower pan mounting screws are often much thinner than lid mounting screws and tend to have flat heads as a matter of design in order to avoid scratching the table top the VCR is sitting on.

Once the lid to your VCR has been removed, you will see mounting screws which are red in color. These are so colored to signify their importance. Often these particular mounting screws will secure the carriage assembly to the metal chassis of the VCR.

Other mechanical components which make up the typical VCR format, such as idler assemblies, tension bands and break arms are often secured into place with general purpose inner mounting screws.

The very small, fine-threaded mounting screws are often used to secure belt guards and various sensors to the chassis of the VCR. Manufacturers tend to over-tighten this size mounting screw. It is very important not to strip the head on this particular size mounting screw or your repair procedure may be halted.

Now, let's begin to outline the proper disassembly steps in order to put your VCR into its serviceable position. This will give you access to the various parts which make up the typical VCR format in order to clean,

WARNING: All VCRs have a warning label secured to the outside rear panel stating, "TO AVOID THE RISK OF AN ELECTRICAL SHOCK, DO NOT REMOVE TOP COVER." Make a point to remove the power supply cord from any power supply outlet before attempting to service your VCR. If this safety precaution is taken into consideration with every VCR, the risk of an electrical shock should be kept at a minimum or eliminated entirely.

replace or adjust the part or area within your VCR format that may be causing the symptom.

If the format of the VCR that you are currently working on is of the top-loading type, it will be necessary to remove the plastic cassette cover on the carriage assembly first before the VCR lid can be successfully removed.

2.2 Removing the Cassette Cover Guard on Top-Loading VCRs

Top-loading VCRs are slightly different from the front-loading format only in regard to the method of receiving the video cassette. When attempting to remove the lid from the top-loading VCR, it is important to remove the plastic cover guard before doing so.

Begin by accessing the two mounting screws which are located on the top side of the plastic cover guard. Often these particular mounting screws will be hidden from sight under small rubber tabs for aesthetic reasons. These rubber tabs are removable in order to perform this procedure. To access other mounting screws that may be located on either side of the cassette cover guard, it will be necessary to press the Eject command so that the carriage assembly and plastic cover guard is in the up position.

Figure 2-2. *Removing the mounting screws from the top of the carriage assembly.*

Once the carriage assembly has been ejected, be certain to inspect both the left and right side of the plastic cover guard for two more mounting screws. When removing these very small mounting screws, be careful not to lose them in the gears of the open VCR (*Figure 2-2*). Once you are certain that each mounting screw has been removed, simply slide the plastic cassette cover guard forward and lift, and it should lift right off of the VCR.

With the plastic cassette cover guard removed, depress the carriage assembly until it locks in place. This is the position the carriage assembly should be in when removing the VCR's lid.

At this point, continue with the disassembly steps as if the VCR you are working on is of the front-loading format.

2.3 Removing the VCR's Lid Mounting Screws

The mounting screws that secure the lid to the plastic frame of the VCR are often the largest mounting screws you will remove. Although similar-size mounting screws secure the metal chassis to the plastic frame, chassis mounting screws are almost never removed during a standard cleaning or repair procedure.

Lid mounting screws are located on five of the six sides of the VCR, depending on the make and model you are currently working on (*Figure 2-3*). With the exception of the front panel, lid mounting screws can be located on the underside, top, rear, and both sides of the VCR.

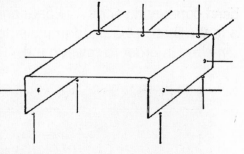

Figure 2-3. Lid mounting screw locations on a VCR.

Keep in mind that these particular disassembly steps can be applied to any VCR model on the market today. Always remember to proceed slowly while looking for the proper screws that secure the lid to the frame of the VCR. Be certain not to use a lot of force when attempting to remove the lid from the body of the VCR. If a hidden mounting screw goes unnoticed, damage may occur if the lid is made of plastic.

2.3.A Top Lid Mounting Screws

Top lid mounting screws are either located along the back leading edge of the VCR or hidden under a "tuner box" door. Earlier models of VCRs often have a tuner box located on top of the VCR for tuning in

WARNING: Once the lid to your VCR is ready to be removed from its plastic frame, refer to Section 2.4, where the proper method for removing the lid from the chassis of your VCR is outlined. If, however, the lid to your VCR is still secured into place, please continue with the disassembly steps.

WARNING: Once the lid of the VCR is removed, DO NOT place bare metal tools directly on the PC board or the capacitors may discharge electricity, resulting in an electrical shock.

specific UHF and VHF television broadcast stations through the VCR. This is an area where a hidden lid mounting screw might be located.

2.3.B Side Mounting Screws

Side mounting screws are common to nearly every VCR on the market today. These mounting screws are most likely the largest that you will remove from the VCR during a standard repair or cleaning procedure. Simply remove these screws and set them aside in a sandwich bag labeled "Lid Mounting Screws".

2.3.C Rear Lid Mounting Screws

Rear lid mounting screws are located at the back of the VCR, either on each corner or running along the upper edge of the VCR's cover. Be certain to look in the dark recesses of the back panel for these particular mounting screws. Some VCR models have a single mounting screw hidden from sight that may hinder the removal process. Once the proper rear mounting screws have been removed, place them in the sandwich bag labeled "Lid Mounting Screws".

2.3.D Lower Pan Lid Mounting Screws

Make a point to check the outer perimeter of the lower pan of the VCR for lid mounting screws. In the VCR models which incorporate lower pan lid mounting screws, Panasonic in particular, an indentation of an arrow will often be used to indicate lower pan mounting screws to be removed in order to release the VCR's top cover.

Figure 2-4. Carefully lift up on the VCR lid to remove it.

Figure 2-5. Examples of foreign objects commonly found in VCRs.

Once you have successfully located the proper lower pan lid mounting screws, go ahead and remove them. Put these mounting screws in the same sandwich bag with the other lid mounting screws which you have removed up to this point.

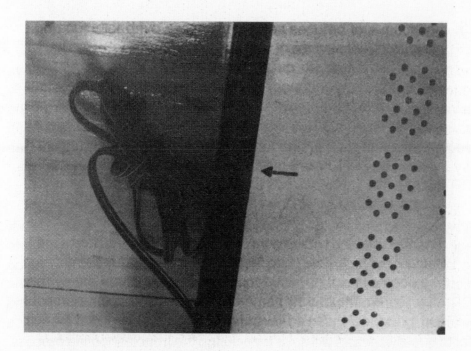

Figure 2-6. Orientation arrow on the lower pan of the VCR.

WARNING: Since you are attempting to locate the cause of the symptom in the VCR, it is NOT recommended that the internal mechanical parts of the VCR format be sprayed with compressed air at any time during the disassembly procedure. If a simple tension spring has come undone through normal use, and happens to go unnoticed for the time being, it does not make sense to spray it out of the body of the VCR and onto the floor where it would certainly be lost for good.

WARNING: Before continuing to the actual lid removal step, it is important to note that many Panasonic brand VCRs use an interlocking tab system which assists in keeping the lid of the VCR flush to its plastic frame. Proceed slowly when removing the VCR's lid to ensure that a locking "foot" is not broken off. If a locking foot is broken off, the VCR lid will not sit flush to the body of the VCR.

Turn the VCR over so it is sitting right side up on your work table and continue with the proper lid removal procedure. It is a good idea to read the WARNING first before attempting to remove the VCR lid, to ensure that damage does not occur to the VCR frame.

2.4 Removing the VCR Lid from its Plastic Frame

After the necessary lid mounting screws have been successfully removed, stand over the VCR and hold the lid with both hands towards the back. Begin gently pulling back until the leading edge of the VCR lid releases from beneath the plastic edge of the front panel. Once the VCR lid has moved several inches towards the back of the VCR, slowly lift up on the lid until it clears the frame of the VCR (*Figure 2-4*). Set aside the VCR lid in a safe place away from your work area, with its proper lid mounting screws, until it is time to reassemble the VCR.

At this point, it is very important to keep foreign objects, including fingers, out of the exposed VCR.

Begin the troubleshooting process by actively looking for the cause of the original fault affecting the VCR. Keep in mind, from this point on, that the process of further disassembling the VCR is now a judgment call. That is to say, removing the lid of the VCR may be all that is required in order to remove dirt buildup, access a mechanical part for adjustment, or remove a foreign object that may be causing the fault.

2.5 Removing the VCR's Lower Pan

The next step in putting a VCR into its serviceable position is to remove the lower metal pan. When the lower pan is removed, you can really actively search for signs of error such as a small toy stuck within the gears, an unhooked tension spring, or a broken drive belt (*Figure 2-5*). Keep in mind, if the faulty part is accessible, removing the lower pan may be all that is required in order to remedy the fault affecting your VCR.

With the VCR on its back, to access the lower pan mounting screws, begin by drawing an "orientation arrow" on the lower pan pointing to the back of the VCR (*Figure 2-6*). This orientation arrow should be located on the lower edge of the lower pan pointing to the back of the VCR. The arrow will denote which end of the lower pan is mounted toward the rear of the VCR. A simple marking such as this will reduce a lot of guesswork and will save time during the troubleshooting process, when the lower pan is held temporarily in a position to protect the various lower side drive gears.

*Figure 2-7. The VCR's lower pan can be lifted off once its
mounting screws and locking tabs are removed.*

In most cases, the lower pan will be secured to the plastic frame of the
VCR with either four or six mounting screws. It is very important to
return each mounting screw to its proper location on the lower pan.
For this reason, it is a good idea to mark each particular mounting
screw with a hash mark (/) before removing them.

Once the correct lower pan mounting screws have been marked and
the orientation arrow has been drawn, remove each mounting screw
and place them in a small sandwich bag labeled "Lower Pan Mounting
Screws".

As you are removing each pan mounting screw, it may be necessary to
make a note to yourself on differences in size, shape or color. If
differences occur, make a note to yourself on the lower pan by utilizing
the various symbols, as outlined in section 1.10.G of Chapter One.

After the proper lower pan mounting screws have been removed, lift
up on the leading edge of the lower pan and it should release from the
plastic frame of the VCR (*Figure 2-7*). If there is any resistance, do not
force it or continue to lift up on the lower pan. On some models of
VCRs, a nylon locking tab is used to secure the lower pan to the frame
of the VCR. Also, the lower pan may slide under the front panel on a
plastic hook as a matter of design. Be certain to search for these
potential drawbacks and retry the lower pan removal step.

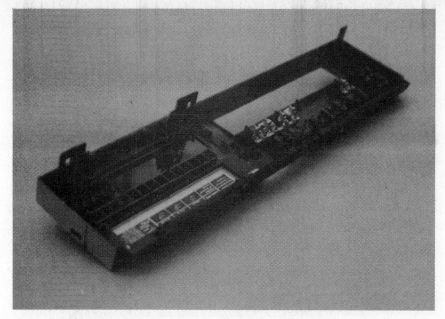

Figure 2-8. The removed front panel of a VCR.

2.6 Removing the VCR's Front Panel

The next step in the disassembly process of the VCR is to remove the front panel. Keep in mind that 99% of all VCR front panels are designed to be removed (*Figure 2-8*).

Although the front panel on a VCR may seem to contain a digital clock display, various command buttons and timer/program switches, often this is not the case. The VCR's front panel is simply an intricately labeled and somewhat intimidating plastic cover.

Manufacturers of VCRs use a plastic locking tab system which keeps the front panel secured to the plastic frame of the VCR. Locking tabs can be located on every side of the front panel along with a potential mounting screw (*Figure 2-9*).

Let's consider the correct steps in removing the front panel of the typical VCR. It is a good idea to use a labeled sandwich bag for tracking control knobs and mounting screws that may be removed.

The first step when removing the VCR's front panel is to locate and remove any potential mounting screws. Be certain to look for a single mounting screw located under the front panel door.

Once the mounting screws have been removed, use light finger pressure to release each of the locking tabs. Do not use a lot of force or the locking tabs may break off. If this occurs, the front panel may no longer

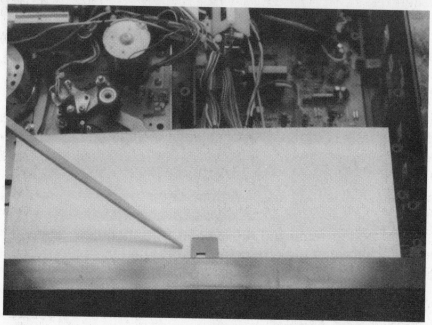

Figure 2-9. A locking tab on a VCR front panel.

sit flush to the command buttons behind it and a symptom may result. After each locking tab has been released, the front panel should lift off from the plastic frame of the VCR.

Panasonic brand VCRs often incorporate a very convenient hinge system that requires the release of only three topside locking tabs before the front panel literally swings down and is removed. After removing the front panels of several different VCR brands, you will find this hinge system most convenient.

Be certain to set aside the front panel to your VCR in a safe place with its mounting screws until it is time to reassemble and test the performance of your VCR.

2.7 Removing a VCR's Head Shield

Many VCRs on the market today contain a head shield as a matter of design. In most cases, the head shield needs to be removed in order to continue with the next disassembly step.

The VCR head shield is either secured to the carriage assembly or to the frame of the VCR with two or more mounting screws. These particular mounting screws are often red in color to signify their importance.

Also, a VCR head shield will either be made of metal, or a PC board within the VCR will be incorporated to cover the upper head assembly that contains the sensitive video heads. Of the two types of head shields, let's consider the removal steps for the metal type first.

2.7.A Removing a Metal Head Shield

Metal head shields are molded to fit into place, and are either mounted onto the chassis of the VCR or secured to the carriage assembly with several mounting screws.

Metal head shields may also have one or more ground wires secured to them. Be certain to mark the location of any ground wires with a "G" for ground before removing the ground wire mounting screw. Your mark will ensure that the ground wire is returned to its proper location when it is time to reassemble your VCR for testing.

Another aspect to take into consideration when removing the metal head shield are plastic locking tabs that help secure the head shield into place. It is required that these locking tabs be released before the head shield can be removed.

Once the location of any ground wires have been marked, remove the mounting screws that secure the metal head shield in place. Release any potential plastic locking tabs and gently lift up on the head shield. It should release from the chassis of the VCR (*Figure 2-10*). Set aside the metal head shield, along with its mounting screws, in a safe place away from your work area.

Figure 2-10. *Removing a metal head shield.*

Figure 2-11. *Removing a PC board head shield.*

2.7.B Removing a PC Board Head Shield

As a matter of design, manufacturers of VCRs often incorporate the use of an important PC board as a head shield. Chances are this PC board is mounted on nylon hinges which allows the head shield to be lifted in order to access a faulty part or to proceed to the next disassembly step. Also, one or more nylon locking tabs may protrude through the surface of this PC board. Do not attempt to lift up on the PC board head shield until the nylon locking tabs are released.

Begin by removing the mounting screws which secure the head shield to the VCR. Often there will be two mounting screws. Release any plastic locking tabs which may protrude through the surface of the PC board or around its perimeter.

Gently lift up on the PC board head shield (*Figure 2-11*). If there is any resistance, STOP and inspect this area for a hooked wire or a ribbon cable that may be hindering your work. Retry and lift up again on the PC board head shield. It should lift easily into its serviceable position. It is a good idea to use a wooden dowel or a large cotton swab to keep the PC board head shield propped up and out of your way.

Chapter Three
REMOVING THE VCR CARRIAGE ASSEMBLY

Chapter Three
REMOVING THE VCR CARRIAGE ASSEMBLY

When a video cassette is inserted into a VCR, the mechanical device that receives the video cassette and lowers it into the VCR is called the carriage assembly. All VCRs on the market today have a carriage assembly as a matter of design. The VCR carriage assembly is an area where faults may occur either through many hours of normal use or as an oversight during a routine cleaning. (*Figure 3-1.*)

Here we begin by outlining the proper steps to remove the top-loading carriage assembly from the VCR.

3.1 The Top-Loading Carriage Assembly

During the manufacturing of the top-loading VCR, the carriage assembly mounting plates were aligned to factory specifications and secured into place with several mounting screws. If the mounting screws are arbitrarily loosened and the mounting plates on either side of the carriage assembly are allowed to shift either slightly forward or backward, the carriage assembly will twist, resulting in a symptom.

Figure 3-1. *VCR carriage assemblies. A top-loading carriage assembly is shown on the left, and a front-loading carriage assembly is on the right.*

Figure 3-2. *A top-loading carriage assembly mounting plate.*

In a top-loading VCR whose carriage assembly is out of alignment with the rest of the VCR's tape path, the cassette housing usually will not stay down when a cassette is entered and the housing is depressed. Also, the tape may "rattle" in the Rewind mode and exhibit a sluggish Fast-Forward mode.

When attempting to extract a top-loading carriage assembly, it is a good idea to etch the perimeter of each mounting plate while the carriage assembly is still mounted within the VCR. Use a small knife or a sharp pencil for this procedure. The carriage assembly mounting plates, which are secured to the chassis of the VCR, are often held in place by four mounting screws (*Figure 3-2*). The outline of each mounting plate that you etched will act as a "stencil" or guide to ensure that the carriage assembly is aligned to its original factory specifications when you are ready to reassemble.

In addition to etching the perimeter of each base mounting plate on the carriage assembly, there are several key aspects or precautions to take into consideration when attempting to remove the carriage assembly from any top-loading VCR.

First, consider the cassette read/record switch, which is most often mounted on the chassis near the lower left corner of the carriage assembly. Be certain not to break the tab off of this switch when attempting to remove or return the carriage assembly during a repair or cleaning procedure.

The second aspect to consider is the mechanical latch/pin system used in the VCR in order to release the carriage assembly when the Eject button has been depressed. Often there will be a lever and latch/pin located just behind the Eject button inside the VCR. It is a good idea to eject the carriage assembly several times in order to study how the lever and latch/pin system is designed. This will assist in returning the latch/pin to its correct position when reinstalling the carriage assembly back into the VCR, without bending or breaking an integral part of this mechanism.

3.2 Removing the Top-Loading Carriage Assembly

Once you have taken these items into consideration and have outlined the base of each mounting plate, proceed to remove the top-loading carriage assembly by first depressing the Eject button. With the carriage assembly in the full eject position, the small latch pins on either side of the carriage assembly are released. This prevents the pins from bending and the carriage assembly from twisting once the base mounting screws have been removed.

Continue the removal process by removing each mounting screw located on the upper and lower mounting plate on either side of the carriage assembly. Place the mounting screws in a sandwich bag labeled "Carriage Assembly". Once all of the mounting screws have been removed, the carriage assembly is free to be lifted from the VCR.

Gently lift the carriage assembly upward to ensure that the cassette read/record switch and the latch/pin system are not broken during the removal process.

After the top-loading carriage assembly has been removed, continue with the descriptions of the various tape path parts and their cleaning procedures as you would if you were working on a front-loading VCR.

3.3 The Front-Loading Carriage Assembly

VCRs that contain a front-loading carriage assembly certainly outnumber the top-loading VCRs on the market today (*Figure 3-3*). Often the removal process for the front-loading carriage assembly is much more straightforward and requires little preparation as opposed to the top-loading type.

In most cases, there are two inner base mounting screws which secure the carriage assembly to the metal chassis of the VCR. These carriage assembly base mounting screws are often red in color to signify their importance.

Figure 3-3. *A typical front-loading carriage assembly.*

The location of the carriage assembly mounting screws will vary from VCR to VCR. However, the two most common locations for these red mounting screws are:
1. At the base of each inner wall of the carriage assembly, or;
2. Along the outer perimeter of the carriage assembly.

Be certain not to confuse the outer perimeter carriage assembly mounting screws with the screws that are used to hold the carriage assembly together. Accidentally disassembling the carriage assembly at this time will often result in a hodgepodge of misaligned gears and a VCR that does not receive a video cassette properly.

Another aspect to take into consideration when attempting to remove the front-loading carriage assembly are ground wires. It is a good idea to mark the location of any connecting ground wires with a "G" on the carriage assembly before disconnecting the ground wire itself. Your markings will assure that the ground wire is returned to its proper location once the carriage assembly has been reinstalled.

As a matter of design, a carriage assembly may have base locking feet which slide into grooves or notches cut into the chassis for added stability. Other models of VCRs may have posts which sit in cups or U-shaped feet hooked under a mounting post located on the chassis. When attempting to remove the carriage assembly from the VCR, DO NOT use force to dislodge this type of locking system. Be careful not to bend or break the locking feet.

When it comes time to reinstall the carriage assembly into the VCR, you must be careful that the bottom of the carriage assembly is not seated on a wire or lying on some other object. For example, if a locking foot on the bottom of the carriage assembly is not inserted properly into its notch in the chassis, the carriage assembly would be ajar in relationship with the other VCR tape path parts. If the performance of the VCR is tested with the carriage assembly in this position, the videotape will load through the VCR's tape path at a slight angle. Besides the frustration created by having to disassemble the VCR in order to correct this human error, the edges on the passing video tape would most likely be ruined, which in turn may damage the VCR's sensitive video heads.

3.4 Removing the Front-Loading Carriage Assembly

The first step toward removing the front-loading carriage assembly is to remove any mounting screws that may be located along the upper left and right edges of the carriage assembly. These particular perimeter mounting screws tend to be red in color to signify their importance.

Located on both the inner left and right side of the VCR's carriage assembly are two mounting screws that are often red in color. They will have to be removed. These particular inner base mounting screws have much tighter threading than the perimeter mounting screws removed in the previous step.

Figure 3-4. Removing the front-loading carriage assembly.

Once the necessary carriage assembly mounting screws have been removed, place them in the sandwich bag labeled "Carriage Assembly" and set it aside in a safe place away from your work area.

At this point the carriage assembly may be lifted from the VCR (*Figure 3-4*). You may think the VCR looks as if it will never be returned to its proper working order once the carriage assembly is extracted, but this is often not the case. Manufacturers of VCRs have designed the carriage assembly to be removed in order to access a dirty or faulty part in the VCR's tape path.

Hold the carriage assembly firmly with one hand and shift it slightly forward while lifting up at the same time. Be certain not to cut yourself on the sharp, unfinished edges. The carriage assembly should lift out from the body of the VCR. If there is any resistance, STOP! Inspect this area of the VCR and proceed slowly. A single mounting screw may have gone unnoticed or a locking foot may still be secured in place at the base of the assembly.

Study how the carriage assembly is secured within the VCR. Do not use a lot of force but try to dislodge it if it seems stuck. The carriage assembly should free itself and will lift out of the VCR after only a few removal attempts.

Once the carriage assembly has been removed from the VCR, disconnect the ribbon or harness cable located at the rear corner of the assembly. This ribbon or harness cable is responsible for providing electricity to the various components in the carriage assembly. There are two distinct methods for releasing this particular harness or ribbon cable:

1. A harness cable will either have a male or female connector. The female of the connector simply unfastens from the male connector that is mounted on the carriage assembly.
2. A ribbon cable will be inserted into a connector that is permanently mounted on the side of the carriage assembly. In order to release the ribbon cable, use a small screwdriver and depress one side of the connector; the ribbon cable should lift off of the connector easily. It is a good idea to identify the leading edge of the ribbon cable with a mark to ensure that the lead wires are reinserted in the correct order.

3.5 Parts of the VCR Carriage Assembly & What They Do

Assuming that all of the VCR disassembly steps have gone well to this point, the carriage assembly should be removed from the VCR. Set aside the assembly in a safe place away from your work area.

WARNING: The leading edges of the carriage assembly are often unfinished and are VERY sharp! DO NOT use a lot of force at this point, to insure that you do not cut yourself.

WARNING: Located at the right corner of the assembly is either a harness or ribbon cable which connects the carriage assembly to the electrical system of the VCR. Use very little force when attempting to disconnect this cable. In some cases, this harness or ribbon cable will be permanently soldered into place. DO NOT attempt to desolder the lead wires. Once the carriage assembly has been removed from the VCR, place it on a piece of cardboard resting on the VCR. Often the various parts of the VCR's tape path can still be accessed for either cleaning or replacement with the carriage assembly in this position. NEVER PLUG THE VCR's POWER SUPPLY CORD INTO A WALL OUTLET WITH THE CARRIAGE ASSEMBLY IN THIS PRECARIOUS POSITION.

The typical carriage assembly in a front-loading VCR contains nine specific "hot spots" which should be suspected to fail without notice:

1. The connecting ribbon or harness cable.
2. The cassette door release tab.
3. Various loading switches.
4. The VCR's loading motor.
5. The connecting loading mechanism; belt, gear, and link.
6. Various loading cam gears.
7. The Cassette-In switch.
8. The Read/Record switch.
9. The End/Start sensors.

A tenth location to suspect a fault in relation to the VCR's carriage assembly is the light-emitting diode or incandescent lamp, hereafter referred to as an LED/lamp. The LED/lamp information has been included at the end of this chapter of carriage assembly parts, since it ties in directly with the overall performance of the VCR's carriage assembly.

Let's now consider the different parts housed in the typical VCR carriage assembly and their symptoms when these particular parts either become dirty through normal use or fail.

3.6 The Connecting Harness or Ribbon Cable

Carriage assemblies are connected to other components within the VCR via a harness or ribbon cable (*Figure 3-5*). Most VCRs on the market today are designed with the ribbon cable connector mounted on the lower right rear corner of the carriage assembly. This ribbon cable transfers vital information from the carriage assembly motor, sensors, and switches to the VCR microprocessor. Be certain not to leave this ribbon cable disconnected after servicing your VCR. If there is a loose connection or break in the flow of electrical information from the carriage assembly to the VCR's microprocessor, the microprocessor notes this as a serious problem and the VCR will exhibit a symptom.

Symptom: *Symptoms of a harness or ribbon cable that is broken or has been left disconnected include: the VCR will not accept a video cassette; the VCR may power up but will not receive a video cassette; or the drive motors are heard rotating when the VCR is powered on.*

Figure 3-5. *Drawings of a connecting harness cable (left) and a ribbon cable (right).*

WARNING: Unless you have purchased a junker VCR to experiment on, do not attempt to disassemble the carriage assembly out of curiosity and assume that all of your "tweaking" will correct an original faulty symptom. The gears located on either side of the carriage assembly were aligned to manufacturer specifications when the VCR was being assembled. Misguided tinkering and "tweaking" often results in a hodgepodge of loose fittings, misaligned gears and missing tension springs. The result is a carriage assembly that once worked well but now refuses to accept a video cassette.

3.7 Cassette Door Release Tab

Located on the right side of the inner wall of the typical carriage assembly is a small plastic or nylon cassette door release tab (*Figure 3-6*). As a matter of design, all video cassettes have a small plastic button located on the right corner of the video cassette shell where the cassette door is hinged. When this plastic button is depressed, the video cassette door can then be lifted and the video tape accessed. When a video cassette is entered into the carriage assembly, the cassette door release tab depresses the plastic button. As the video cassette is gently lowered into the VCR, the door to the video cassette is opened so that the videotape can be extracted and threaded through the VCR's tape path.

Figure 3-6. Drawing of a cassette release tab.

The cassette door release tab can be the cause of a symptom. Many VCR cassette door release tabs have a tension spring that keeps the release tab in contact with the video cassette as it enters the carriage assembly. If this tension spring or release tab jams, breaks or sticks, the symptom might be:

1. The video cassette enters the carriage assembly at a slight angle and gets stuck.
2. The door to the video cassette does not open and the VCR quickly ejects the video cassette.

There may also be problems with the VCR's cassette-loading switch. When a video cassette enters into a VCR, the cassette-loading switch is activated. The cassette-loading switch can be located in several different places on the carriage assembly, depending on the make and model of VCR. Often, cassette-loading switches will be incorporated in the loading gears mounted on either side of the carriage assembly. When a cassette-loading switch is activated, an electrical signal is sent to the VCR's microprocessor. This signal informs the microprocessor that a video cassette is entering the VCR's carriage assembly. After receiving this initial signal, the microprocessor emits another signal to the VCR's carriage assembly activating the loading motor in order to receive the video cassette for Play or Record use.

Symptom: *If a cassette-loading switch is defective, or the contacts are dirty or bent in any way, the proper electrical signal will not be emitted to the microprocessor and the VCR may refuse to receive the video cassette being entered into the VCR. Be certain to suspect a dirty or bent cassette-loading switch if your VCR powers up okay but refuses to accept or eject a video cassette.*

3.8 The Cassette-Loading Motor

After the cassette-loading switch has been activated, a signal to turn on the carriage assembly loading motor (*Figure 3-7*) is sent out from the VCR's microprocessor. The loading motor receives this electrical signal and begins rotating in the proper direction to lower the video cassette into the VCR. When the Eject command has been selected from the front panel of the VCR, the VCR's microprocessor interprets this signal and sends out the appropriate signal to the loading motor instructing it to begin rotating in the opposite direction in order to eject the video cassette.

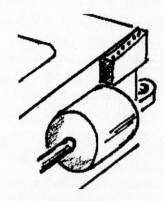

Figure 3-7. *Drawing of a cassette-loading motor.*

Symptom: *Loading motors are subject to fail at any time during the VCR's useful life. If a loading motor becomes very hot while the VCR is plugged in or if the shaft of the loading motor stops rotating, chances are the loading motor is faulty. If this is the case, the VCR may refuse to either accept or eject a video cassette. A loading motor may fail due to:*

1. *Power surge in the VCR power supply cord.*
2. *A too-tight newly installed drive or loading belt.*
3. *Wear and tear from many years of normal use.*

After many years of receiving and ejecting video cassettes, these small motors have a tendency to overheat, freeze up, and eventually stop rotating. Always suspect a failed loading motor if the carriage assembly suddenly refuses to receive or eject a video cassette. A simple test can be performed with a nine volt battery to see whether or not the loading motor is capable of receiving and ejecting a video cassette. See Chapter One, Section 1.10.F, *The Nine-Volt Jolt.*

3.9 Three Common Loading Motor Connecting Rod Assemblies

Depending on the make and model of the VCR you are currently working on, chances are it will contain one of the three loading motor connecting rod formats outlined here. The different methods that VCR manufacturers use will often fall into one of the following categories:

1. Loading Belt.
2. Worm Gear.
3. Direct Link.

The three different methods of connecting the VCR loading motor to the carriage assembly cam gears is presented in order of their reliability. Of course, each method is susceptible to having the loading motor fail at any time. Also, as a matter of design, one particular method may begin to exhibit a symptom more frequently than another.

3.9.A The Loading Belt Format

The first method used by manufacturers to exhibit a symptom through normal use is the loading belt format. In this case, a small loading belt is used to transfer the torque of the loading motor to the cam gears of the VCR, in order to receive and eject the video cassette (*Figure 3-8*). After many hours of normal use, this loading belt has a tendency to run smooth and stretch, and begins to loose traction.

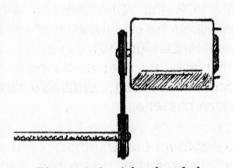

Figure 3-8. *A loading belt connecting rod format.*

Replacement parts can be purchased at your local electronics shop.

Symptom: The symptom for a slipping loading belt might be a loud "squelching" noise emitting from this area each time the VCR is plugged into a wall outlet or when a video cassette is entered into the carriage assembly. Also, if this loading belt breaks, the VCR will neither accept nor eject a video cassette.

3.9.B The Worm Gear Format

In this example, the small loading belt is absent. Instead, manufacturers have placed a worm gear on the shaft of the loading motor in direct contact with the carriage assembly cam gear (*Figure 3-9*). As the video cassette is entered into the VCR, the worm gear rotates in the proper direction via the loading motor. In turn, the connecting cam gear rotates, thus lowering the video cassette into the VCR. This simple, straightforward design may seem to be glitch-free. However, it is susceptible to failure through outside variables, which results in a symptom.

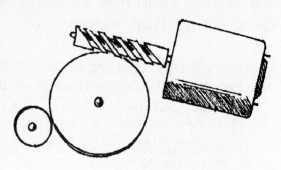

Manufacturers of VCRs use a graphite-based grease on the worm gear teeth which tends to dry out over time and harden. When the hardened

Figure 3-9. A worm gear.

grease accumulates in the threads of the worm gear, the teeth of the connecting cam gear tend to bind and may stop rotating. The hardened grease eventually hinders the performance of the carriage assembly.

Symptom: The symptom for this might be that the VCR refuses to either accept or eject a video cassette. Also, a loud "wrenching" sound may be emitted from the worm gear each time a video cassette is inserted or ejected from the VCR. To remedy this problem, simply apply a generous amount of cleaning solution to the worm gear and regrease its threads.

3.9.C The Direct Link Format

Of the three methods used for transferring the torque of the loading motor to the carriage assembly loading gears, the direct link method tends to be the most reliable and maintenance free. Here, two interlocking nylon clamps join to form a single rotating axle (*Figure 3-10*). This method is most often found in Matsushita (Panasonic) brand VCRs.

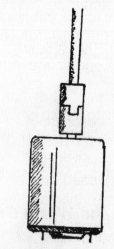

The loading motor can be extracted from the side of the carriage assembly by prying it out once the rear mounting screw has been removed and the nylon locking tab has been released. Be certain not to break the interlocking nylon clamps on each axle rod when attempting to remove the faulty motor. Replacement parts can be purchased at your local electronics shop.

Figure 3-10. A direct-link connecting rod format.

Figure 3-11. *Loading cam gears.*

3.10 Carriage Assembly Loading Cam Gears

When a video cassette is entered into the carriage assembly, often a loading cam gear will trigger a series of On/Off switches which initiate the loading and unloading process (*Figure 3-11*). Loading cam gears are located on either side of the VCR's carriage assembly. When the gear is rotated via the loading motor, a small arm, post or ridge on the cam gear turns the appropriate carriage assembly switch on or off to execute either the load or Eject command.

VCR loading cam gears are often assembled in layers that contain a tightly wound spring at the core. This spring provides proper tension to the entire gear assembly. Tension springs are often held in place with a single nylon tab protruding from one of the gears. These nylon tabs are susceptible to sheering off after many hours of normal use. Whenever a nylon tab breaks off, the carriage assembly will no longer receive a video cassette properly since the inner tension spring has been "released" and no longer provides the necessary tension. If this is the case, the cam gear will need to be replaced in order to restore the carriage assembly to its proper working order. Replacement parts can be purchased at your local electronics shop.

3.11 Cassette-In Switch

Once the video cassette has been successfully entered into the VCR via the carriage assembly, a Cassette-In switch is activated. The

Cassette-In switch is often mounted on the chassis of the VCR. When this particular switch is activated, the VCR microprocessor receives an electrical signal which indicates the video cassette has been successfully seated into the VCR.

Symptom: *The VCR receives a video cassette but refuses to either go into a particular mode as instructed from the front panel, or the VCR immediately ejects a video cassette once the video cassette is seated within the format of the VCR.*

3.12 Read/Record Switch

The last of the various switches to consider on the carriage assembly as the video cassette is being entered into the VCR is the read/record switch. Like the Cassette-In switch, this switch is often mounted on the chassis of the VCR or on the carriage assembly itself. The Read/ Record switch informs the VCR microprocessor whether or not the video cassette can accept recording.

Located at the rear left corner of the typical video cassette is a record safety tab. If this tab is present, the Read/Record switch is activated. The VCR microprocessor notes this and the videotape can be recorded over. On the other hand, if the record safety tab is NOT present, as is the case with most rental movies, the Read/Record switch is NOT activated. The VCR microprocessor notes this and the videotape cannot accept recording; the VCR will only read (play) information from the video cassette.

In the event that a video cassette contains valuable footage which you would like to preserve, simply break off the record safety tab. If someone unknowingly enters the video cassette into their VCR and attempts to record over it, the VCR will most likely eject the video cassette since the microprocessor detects that the safety record tab is missing.

Symptom: *If the contacts of the Read/Record switch are bent or become dirty through many hours of normal use, the VCR may not go into Record mode. Also, the VCR may go into the Record mode from the front panel commands but will not go into Record mode when the VCR's timer is set and activated.*

3.13 End & Start Sensors:
Infra-red Diode/Incandescent Lamp

Located on the lower side of every VHS video cassette is a small hole between the take-up and supply spools. Also, there are two smaller holes which tend to go unnoticed, located on both the left and right sides of the video cassette shell. The side holes can be seen when the door to the video cassette is lifted open. When a video cassette is entered into the VCR, the hole located on the bottom of the cassette shell between the two spools is lowered over an infra-red emitting diode (LED) or an incandescent lamp (lamp), depending on the make and model of your VCR.

The LED/lamp is continuously emitting a beam of light once the VCR is powered on. As a matter of design, this beam of light travels through the body of the video cassette shell and exits the left and right side holes. Also, as a matter of design, the videotape housed within the plastic cassette shell has several inches of clear leader tape at both its beginning and end.

When the VCR is in the Play or Fast-Forward mode, the videotape travels from the left side of the VCR to the right. When the section of clear leader tape is reached, the light emitted from the LED/lamp located under the video cassette shines through the body of the cassette shell, through the clear leader tape, and hits a light-sensitive sensor mounted on the left side of the carriage assembly. The moment light hits the sensor, a signal is emitted to the VCR microprocessor indicating that the videotape has reached its end. Thus, the left-hand sensor is referred to as the END sensor.

The VCR microprocessor interprets this signal and sends out an "all-stop" command as a safety feature to prevent the VCR from snapping the videotape.

When the VCR is either in the Rewind or Review mode, the videotape travels through the VCR from right to left. When the section of clear leader tape is reached, light emitted from the LED/lamp triggers the light-sensitive sensor located on the right side of the carriage assembly. The VCR microprocessor interprets this signal and again sends out an "all-stop" command in order to prevent the VCR from snapping the videotape. Since the videotape has been rewound to the beginning or start, the right-hand sensor is appropriately referred to as the START sensor.

Symptom: *When either an End or Start sensor fails, the capstan motor and idler drive assembly often rotate simultaneously in the same direction for several seconds before the VCR either stops rotating or powers off. The symptom for a failed infra-red diode is often not obvious. In most cases, the VCR may receive a video cassette, but the only command the VCR responds to is the EJECT command.*

It is a good idea to purchase an infra-red light detector, which can be found for less than $14. To use the detector, simply power-on the VCR and hold the detector to the side of the infra-red diode. If the infra-red diode emits a beam of light, the surface of the detector will glow a bright orange.

Older VCRs tend to use incandescent lamps instead of modern infra-red emitting diodes. The infra-red light detector will not work as an indicator for a failed incandescent lamp. Instead, a visual inspection is often all that is required to confirm whether or not the incandescent lamp is working correctly or needs to be replaced.

Symptom: *The symptom for a failed incandescent lamp is the same for a failed infra-red diode. That is to say, the only command the VCR responds to is the EJECT command. Often, when an incandescent lamp fails, it will become burnt gray in color.*

Chapter Four
PARTS OF THE VCR
TAPE PATH:
What They Do &
How to Clean Them

Chapter Four
PARTS OF THE VCR TAPE PATH:
What They Do & How to Clean Them

With the carriage assembly successfully removed from the VCR, the many parts which make up the typical VCR tape path are now in full view and readily accessible for a good cleaning, adjusting, or replacement. (*Figure 4-1.*)

As with each disassembly step, removing the VCR's carriage assembly will become a "judgment call" for each VCR you service in the future. For now, the carriage assembly has been removed in order to illustrate the extent that a VCR can be disassembled to access the tape path parts in order to correct a fault.

The REAL secret to VCR repair and cleaning is to:
1. Proceed slowly while actively searching for what may be causing the original fault.
2. Learn the function of the individual parts which make up the typical VCR mechanics.

Figure 4-1. *An open VCR showing the tape path.*

3. Obtain the knowledge to recognize the difference between symptoms when a particular part is dirty or has failed.
4. Use the proper disassembly and reassembly steps outlined in this text to access the tape path to either clean, adjust or replace it.

Aside from the small hand tools discussed in Chapter One, the cleaning materials required for cleaning the VCR's tape path parts and video heads consist of:
1. Cleaning solution (a mild fingernail polish remover works well).
2. Cotton-tipped cleaning sticks or swabs.
3. Leather-tipped video head cleaning sticks which can be purchased at most local electronics shops.

4.1 Parts of the VCR Tape Path

As a matter of design, when a video cassette is entered into the VCR, the carriage assembly gently receives the video cassette and lowers it over two sets of guideposts. When either the Play or Record command is selected from the VCR's front panel, the guideposts are mechanically extended forward to their full V-block position. When this occurs, the videotape is extracted from its plastic cassette shell. When the guideposts are fully extended into the V-block position, it is said that the videotape has been "applied to the VCR's tape path."

4.1.A The Supply Spool
When the video cassette is entered into the VCR, the carriage assembly lowers the video cassette onto both a supply and take-up spool. The supply spool is located on the left of the two spools, and the take-up spool is on the right.

The supply spool is where the videotape will begin its trek out of the cassette shell and through the VCR's tape path, to be collected on the take-up spool. (*Figure 4-2.*) Considering the only function of the supply spool is to rotate on its mounting post, this particular part rarely needs cleaning. However, on older VCRs, a rubber drive tire may be located around the perimeter of the supply spool for added traction. If this drive tire runs smooth and looses traction, a symptom may occur.

Symptom: *If the VCR you are working on does have a rubber drive tire located around the perimeter of the supply spool, and the drive tire has worn smooth, the symptom might be that the VCR exhibits a sluggish rewind.*

WARNING: The same procedures for cleaning the VCR tape path parts should NOT be applied to the video heads. That is to say, a foam or cotton-tipped swab should NEVER be used to clean the video heads. The cotton fibers or the porous foam tip can easily snag one of the video heads and chip it off, rendering the video heads useless.

Figure 4-2. *The tape path supply spool.*

Cleaning Solution: To remedy this problem, it may be necessary to remove the supply spool and replace the faulty drive tire. Replacing drive tires is outlined extensively in Chapter Five. Otherwise, simply inspect the perimeter of the supply spool to ensure a foreign object is not inhibiting its performance.

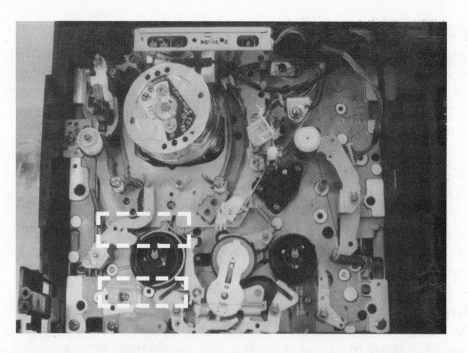

Figure 4-3. *The tape tension band.*

Figure 4-4. The tension arm.

WARNING: *The tape tension band was set to factory specifications by the manufacturer when the VCR was being assembled. Only if the tape tension band shows obvious signs of damage or if the felt pad has dried and fallen off should an attempt be made to remove it. The proper amount of tension applied to the supply spool via the tension band is very important for correct video and audio playback clarity.*

4.1.B The Tape Tension Band

Located around the supply spool is the tape tension band. (*Figure 4-3.*) This band applies proper tension to the supply spool while the VCR is in use. By applying tension to the supply spool, the tension band ensures that the videotape will not sag or "drift" away from the tape path and video heads while the VCR is in use.

Symptom: The tape tension band often becomes very dry through many hours of normal use. The white felt pad may separate from its metal band and fall off completely. If this is the case, the audio and video portion of a playing video tape may become greatly distorted. Horizontal lines may appear on the playback picture as well. Adjusting the tracking control knob often does not help.

Cleaning Solution: To remedy this problem, it may be possible to use a strong bonding cement or glue to secure the white felt pad onto its metal band. If the felt pad is ruined, the tension band will need to be replaced with a new one before the VCR will exhibit a clear picture in both the playback and recording modes.

4.1.C Tension Arm

When the video tape is extracted from its plastic cassette shell during Play or Record mode, a tension arm is mechanically maneuvered into position against the passing videotape. The tape tension band is secured to the base of the tension arm, which assists in keeping the passing videotape stable and jitter free while the VCR is in use. (*Figure*

4-4.) This particular tension arm also has a tension spring located at its base. If this tension spring has come off or is left unhooked after a cleaning or repair procedure, the video and/or audio portion of a playing videotape may be dramatically effected.

Symptom: *The symptom for a tension arm spring that has been left unhooked might be lines of video "noise" running horizontally across the playback screen. Also, the audio portion to the videotape may exhibit an "underwater" or alien sound quality to it. Adjusting the tracking control knob often does not remedy the problem.*

Cleaning Solution: Clean the leading edge of the tension arm post that comes in contact with the passing videotape with a swab dipped in cleaning solution. Use an up-and-down motion to remove accumulated dust and dirt deposits. Visually inspect the base tension spring to ensure that it is secured in place.

4.1.D *Impedance Rollers*

Manufacturers may incorporate several impedance rollers throughout the VCR's tape path, depending on the make and model of the VCR. (*Figure 4-5.*) Impedance rollers in today's VCRs are constructed of nylon, while in older models the material of choice was brass.

Impedance rollers assist in keeping the passing videotape stable and jitter free while the VCR is in use. The impedance rollers were set to factory height specification when the VCR was assembled and should not be randomly adjusted.

Figure 4-5. *An impedance roller.*

Symptom*: If an impedance roller has collected dust, dirt deposits, or pet or human hair, the playback picture may "jitter" uncontrollably. Also, if the impedance roller sticks or stops rotating, a loud "squelching" sound may emit from within the VCR while in the Play, Record, or Review mode.*

Cleaning Solution*:* Clean the nylon or brass impedance roller with a cotton or foam-tipped swab dipped in cleaning solution. Be certain to rotate the impedance roller to ensure the that entire surface is cleaned. Also, make a point to visually inspect each impedance roller to ensure they are both dirt and oil-free. Something as simple as a pet or human hair wrapped around an impedance roller may create a "jitter" in the playback video picture.

4.1.E Full Erase Head

In most cases, unless it is a video cassette player (VCP), the VCR will have a full erase head located directly after the impedance roller. (*Figure 4-6.*) The purpose of the full erase head is to erase all previous video information from the passing videotape while the VCR is in Record mode. Video cassette players, or VCPs, do not require a full erase head in their format since they do not contain the necessary circuits to record onto videotape.

Since the full erase head plays such an integral role in the VCR's erasing and recording process, be certain not to scratch the surface of this head.

Figure 4-6*. The full erase head.*

Figure 4-7. *The guideposts.*

Symptom: *Through many hours of normal use, the leading edge of the full erase head tends to become a magnet for rug fibers, lint, pet hairs, and dirt deposits. One particular aspect of the full erase head that has a tendency toward errors is the connector at the top of the head. The lead wires of this connector may become loose, thus resulting in a symptom. The symptom for loose lead wires at the full erase head connector might be that PREVIOUS video and/or audio information is still present on a new recording. If this is the case, chances are the full erase head connector is the culprit.*

Cleaning Solution: To remedy this "previous information" recording problem, it may be necessary to snip off the full erase head connector and solder each lead wire back into place. Otherwise, simply clean the full erase head with a cotton or foam-tipped swab dipped in cleaning solution. Use an up-and-down motion to remove dust, pet hairs and dirt deposits.

4.1.F The Guideposts

When a video cassette is entered into a VCR, the videotape is lowered over a left and right set of guideposts. (*Figure 4-7*.) The guideposts are mechanically extended forward when the Play or Record command is selected from the front panel of the VCR. When this occurs, the videotape is extracted from its plastic cassette shell and applied to the VCR's tape path.

Each guidepost consists of a stationary arm set at a slight angle, and an adjustable guidepost that contains a free-spinning nylon collar. An adjustment nut is located at the top of each guidepost, which regulates its height.

Also located at the base of each guidepost is a locking nut. This locking nut prevents the entire guidepost from rotating while the VCR is in use. Do not loosen the top adjustment nut or base locking nut out of curiosity or a loss of both video and audio may result.

Instructions for adjusting each guidepost are outlined in more detail in Chapter Seven.

Symptom: If a pet or human hair wraps around either guidepost, the faulty symptom might be that the playback picture "jitters" or randomly fluctuates. Be certain that each guidepost is dirt and oil free.

Often, in NEC brand VCRs, the entire guidepost may loosen and spin freely within the mounting plate that runs within the guidepost groove. If this is the case, the height of the loose guidepost will "drift" while the VCR is in use. The symptom for a loose guidepost might be horizontal lines that appear on the playback screen. Often, adjusting the tracking control knob will not correct the problem until the guidepost is secured in its mounting plate at the correct height.

Cleaning Solution: Use a cotton or foam-tipped cleaning swab dipped in cleaning solution to clean both the stationary arm and the nylon collar on each guidepost. Be certain to rotate the nylon collar on the larger guidepost to ensure all dirt deposits, rug fibers and pet hairs have been removed.

4.1.G The Guidepost Grooves

When the guideposts are mechanically positioned forward to extract the videotape from its plastic cassette shell, each guidepost runs within a groove cut into the chassis of the VCR. These guidepost grooves should be suspected of causing some symptoms. (*Figure 4-8.*)

Manufacturers of VCRs often use a graphite-based grease on each guidepost groove. Through many hours of normal use, the grease tends to dry and harden. When this occurs, the guideposts within these grooves may become so embedded with hardened grease that they have difficulty reaching their fully extended V-block position.

Also, if a piece of popcorn or another alien object enters the VCR and lodges itself within the V-block at the end of either guidepost groove, a symptom may occur.

WARNING: Do not "tweak" the height adjustment nut, located at the top of each guidepost, out of curiosity. EACH GUIDEPOST HAS BEEN SET TO FACTORY HEIGHT SPECIFICATIONS. If the height of a guidepost is adjusted in an attempt to correct a fault, the video and audio portion of a playing videotape may become PERMANENTLY distorted. If this is the case, it may be necessary for an experienced VCR service center to correct the problem.

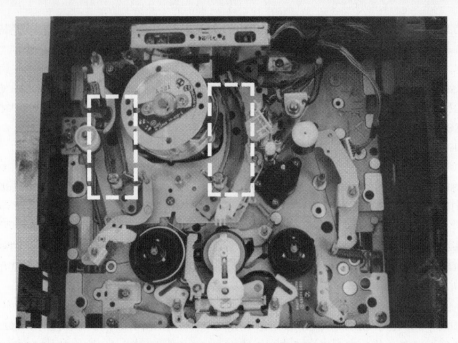

Figure 4-8. Guidepost grooves.

Symptom: *If the guideposts are inhibited in any way from reaching their fully extended V-block position, either from hardened grease or a foreign object lodged within the V-Block, the faulty symptom might be "snow" or horizontal lines that appear on the television screen when the VCR is playing a videotape.*

Cleaning Solution: To remedy this problem, clean each guidepost groove with a cotton swab dipped in cleaning solution. Remove hardened grease that may be inhibiting the performance of each guidepost. It is a good idea to apply new VCR grease to each guidepost groove as well. Also, inspect each V-Block, located at the end of each guidepost groove, for foreign objects that may hinder the travel of each guidepost.

4.1.H The Audio/Control Erase Head Assembly

Once the videotape has crossed the right guidepost, it comes in contact with an audio/control erase head assembly, or ACE head. (*Figure 4-9.*) This assembly is a configuration of two different heads.

The first head the videotape encounters in the audio erase head. The audio erase head erases all previous audio information from the passing videotape while the VCR is in Record mode. When the VCR is in Play mode, the audio erase head is turned off.

The second head the videotape encounters consists of two halves, an upper half and a lower half. Let's first consider the upper half of the second head in this assembly.

WARNING: Never attempt to adjust the screws located on the ACE head base plate. These particular adjustment screws were set to factory height specifications when the VCR was assembled. The tiniest amount of random "tweaking" may have a severe effect on both the audio and video portion of a playing videotape.

When a videotape travels through the VCR's tape path while the VCR is in Record mode, the UPPER half of the second head records audio information onto the upper edge of the passing videotape. When the VCR is in Play mode, the UPPER half of the same head reads the audio information from the upper edge of the videotape. This is where the audio portion of a playing videotape originates.

The LOWER half of the second head in this assembly is called the control head. When a VCR is placed in Record mode, the lower half of the second head records coded information onto the lower edge of the passing videotape. The VCR's microprocessor interprets this coded information in order to keep the various motors operating at their correct playback and record speeds.

Symptom: *The symptom of an ACE head that has accumulated dirt, dust and videotape deposits through many hours of normal use might be that the audio portion of a playing videotape will be very faint. Also, the VCR may randomly alternate between play speeds (SP, LP and SLP) if the control head has accumulated dirt or is out of alignment with the rest of the VCR's tape path parts.*

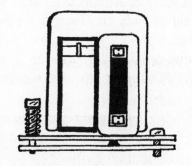

Cleaning Solution: Clean the ACE head with a cotton or foam-tipped swab dipped in cleaning solution. Use a left-to-right motion to clean the affected area of the ACE head, which may be located under a slight overhang. This area plays an integral role in receiving the audio portion of a playing videotape and should be kept dirt and oil free.

4.1.I Stabilizing Guidepost

Once the videotape leaves the ACE head assembly, often it will pass over a small stabilizing guidepost. (*Figure 4-10.*) The purpose of this guidepost is to keep the passing videotape flat against the audio control erase head, as well as provide proper tension to the videotape before it enters the rotating capstan post and pinch roller.

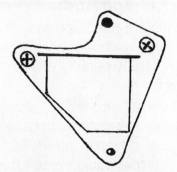

Figure 4-9. A drawing showing the side and top of the ACE head.

Symptom: *Through many hours of normal use, this small stabilizing guidepost may become either too high or too low* *with respect to the other VCR tape path parts. When this occurs, the upper or lower edge of the passing videotape may become "scalloped". If a scalloped videotape is played, the upper or lower portion of the playback picture may flutter or jitter randomly.*

Cleaning Solution: Clean the leading edge of the stabilizing guidepost with a cotton or foam-tipped swab dipped in cleaning solution. Use an up-and-down motion to ensure that all dirt deposits, dust and rug fibers have been removed.

Also, if the upper or lower edge of the passing videotape is being scalloped, it may be necessary to adjust the locking nut on top of the stabilizing guidepost. This adjustment will allow you to either raise or lower the stabilizing guidepost, thus correcting the scalloping.

4.1.J Capstan Post & Pinch Roller

Since the pinch roller and capstan post play such an integral role in the performance of a VCR, it is best to describe their function as single working unit.

After the videotape leaves the stabilizing guidepost, it encounters a rotating capstan post and rubber pinch roller. When the VCR is in either Play or Record mode, the rubber pinch roller is mechanically positioned tightly against the rotating capstan post. The capstan post and pinch roller work as a single unit to pull the videotape through the VCR's tape path. Let's consider the capstan post first:

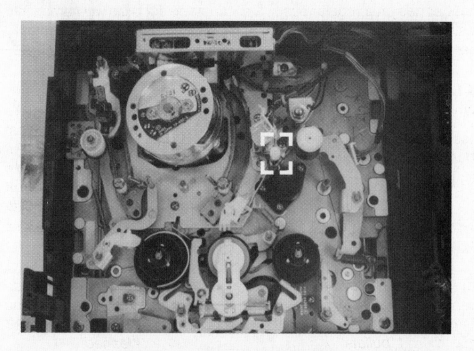

Figure 4-10. The stabilizing guidepost.

Figure 4-11. *The capstan post.*

4.1.J.a Capstan Post

The capstan post is a direct extension of the capstan motor, which is mounted on the lower side of the VCR chassis. (*Figure 4-11.*) It is very important that the surface of the capstan post remain dirt and oil free, to ensure that the VCR continues to perform correctly.

After many hours of normal use, distinct bands of dirt deposits tend to accumulate on the capstan post. If the surface of the capstan post is not cleaned on a regular basis, the videotape deposits will harden into a single band of dirt. When this occurs, the capstan post may eventually damage the surface of any videotape that is played on the VCR.

4.1.J.b Pinch Roller

When either the Play or Record command is selected from the VCR's front panel, the pinch roller is mechanically positioned tightly against the rotating capstan post, thus pulling the videotape through the VCR's tape path. (*Figure 4-12.*) Through many hours of normal use, the rubber pinch roller tends to collect oil, dirt, and videotape deposits. If the surface of the pinch roller is not cleaned on a regular basis, it may become slick and "pitted". If this is the case, the pinch roller will lose traction against the rotating capstan post, resulting in a symptom.

Symptom: *The symptom for a capstan post that has accumulated bands of videotape dirt deposits might be that the surface of the passing videotape becomes pockmarked. The playback picture of this videotape may exhibit "bullets" or snow if the capstan is not cleaned.*

Figure 4-12. The pinch roller.

The symptom for a pinch roller whose surface has run smooth and has become pitted might be a playback picture of a videotape that comes and goes with intermittent patches of snow. Also, an upper or lower edge of the passing videotape may become scalloped. If this is the case, the audio portion of a playing videotape may exhibit an "underwater" or otherwise alien sound quality.

Cleaning Solution: Clean both the capstan post and pinch roller with a cotton or foam-tipped cleaning swab dipped in cleaning solution to loosen and remove dirt. It may be necessary to use a razor blade to dislodge the bands of dirt on the capstan post.

It is recommended that the pinch roller be cleaned with a leather chamois that has been dipped in cleaning solution. Since the pinch roller is notorious for collecting dirt, it would not be uncommon to use ten to fifteen cleaning swabs to clean this VCR tape path part. The leather chamois saves time and ensures that the entire surface of the pinch roller has been cleaned.

4.1.K The Pullout Arm

After the videotape successfully travels between the capstan post and pinch roller, it is often threaded around a small pullout arm. The pullout arm assures that proper tension is applied to the passing videotape, keeping it stable and guiding it back into the plastic shell of the cassette as it is collected onto the take-up spool.

Symptom: *If the height of the pullout arm is out of alignment with the other tape path parts, either the upper or lower edge of the passing videotape may become scalloped. If this is the case, adjust the height adjustment nut located at the top of the pull out arm in order to correct the scalloping problem.*

Cleaning Solution: Clean the leading edge of the pullout arm with a cotton or foam-tipped swab that has been dipped in cleaning solution. Use an up-and-down motion to ensure that all dust, rug fibers, and dirt deposits are removed.

4.1.L The VCR's Take-up Spool

At this point, the videotape has successfully completed its journey through the VCR's tape path. When the video cassette was initially entered into the VCR, the carriage assembly lowered the video cassette over a spool located on the right. Here is where the videotape is collected after its journey. (*Figure 4-13.*)

The take-up spool does not have a tape tension band similar to the supply spool. Instead, a take-up sensor is located underneath the take-up spool. (*Figure 4-14.*) The take-up sensor continuously emits an electrical signal to the bottom of the take-up spool. As a matter of design, a "pinwheel" of reflectors has been painted on the underside of the take-up spool. While the take-up spool rotates on its post when the VCR is in use, the electrical signal emitted by the take-up sensor is interrupted by the rotating pinwheel reflectors. The interrupted reflected

Figure 4-13. *The take-up spool.*

WARNING: *When a VCR is in use, the take-up spool rotates either by a direct set of gears or through the use of an idler drive assembly. The idler drive assembly plays an integral role in the VCR's overall performance. Chapter Five outlines several common idler drive assemblies used in VCRs on the market today. It is a good idea to read Chapter Five before attempting to extract the idler drive assembly from the VCR you are currently working on.*

WARNING: It is very important that the pinwheel design on the lower side of the take-up spool remain reflective. However, a strong cleaning solution SHOULD NOT be used to clean the reflective areas on the pinwheel. USE TAP WATER TO CLEAN THE REFLECTORS!!! If a strong cleaning solution is used, the paint between each reflector may dissolve and the pinwheel design will no longer exist. If this were to happen, the predictable pulse train signal would be inconsistent, resulting in a constant "all-stop" command.

Figure 4-14. The take-up sensor and the bottom of the take-up spool showing the pinwheel design.

signal is called a "pulse train". The VCR microprocessor relies on this pulse train signal to ensure that the take-up spool is rotating at its correct Play or Record speed.

If there is a break in this predictable pulse-train signal, the microprocessor notes this interruption and signals an "all stop" command to the VCR's motors, stopping the advancing videotape in its tracks. This is a built-in safety feature designed to protect the playing videotape from potential damage.

Symptom: If the take-up sensor fails, the symptom might be that the VCR will attempt to play a videotape but immediately defaults to the Stop mode. If the pinwheel on the underside of the take-up spool becomes "foggy" through many hours of normal use, the VCR may randomly stop while playing a videotape. Also, if the digital tape counter does not increment while the VCR is playing a videotape, chances are the take-up sensor is faulty. However, it is possible to rotate the take-up spool with the VCR lid off in order to test whether or not the digital tape counter is incrementing.

Cleaning Solution: Clean the reflectors of the pinwheel on the underside of the take-up spool with a DRY cotton or foam-tipped cleaning stick. If you feel you must use a solution of some sort on the swab, please use water.   DO NOT USE CLEANING SOLUTION IN THIS AREA; if the pinwheel design is distorted, the reflected pulse train signal will not be consistent. If this is the case, the microprocessor may issue an "all stop" command to the motors, and the VCR will stop playing. This is a built-in safety feature to prevent the videotape from further damage.

4.2 Cleaning the VCR Video Heads

Since the video heads are considered the "eyes" of your VCR, it is very important to clean this area of the VCR tape path LAST. This will ensure that any dust that may have been stirred up while cleaning the other tape path parts will be removed from the video head assembly.

Before attempting to clean the video heads, it is vital that you read and understand the following list of NEVER-DOs. The video heads are a very sensitive area of the VCR tape path and should be cleaned properly:

1. NEVER attempt to clean the video heads with the VCR's power supply cord plugged in or while the VCR is playing or recording a videotape. This would certainly result in a more costly repair.

2. NEVER clean the video heads with a cotton or foam-tipped cleaning stick. The cotton fibers or porous foam tip can easily snag a video head and chip it off. Only a manufacturer's authorized leather-tipped video head cleaning stick should be used to clean this area of the VCR tape path.

3. NEVER use an up-and-down motion while cleaning the video heads with the leather-tipped video head cleaning stick. Although other tape path parts require this cleaning motion, the video heads required a left-to-right motion with very little pressure applied.

When cleaning the video heads, the first step is to properly locate them on the upper head assembly of your VCR. Much confusion arises when considering the location of the video heads in a typical VCR. As a matter of design, the upper head assembly is secured to a lower drum motor. The lower drum motor is located within the stationary lower drum assembly. When a VCR is either in Play or

Figure 4-15. *Location of the video head.*

Record mode, the upper head assembly rotates at approximately 1800 rotations per minute. It is this freely-rotating upper head assembly (which contain the video heads) that should be kept dirt and oil free. Also, when the video heads are diagnosed as faulty, it is this upper head assembly that has to be replaced.

4.2.A Step One: Locate the Video Heads in Your VCR

When looking at the upper head assembly, the video heads can be seen as small notches. (*Figure 4-15.*) It is very important not to touch the video heads directly with your fingers. Oil deposits from your fingers can result in severe damage to this sensitive area in the VCR.

Make a point to first rotate the upper head assembly slowly to count the number of video heads in the VCR. Counting the number of video heads will give you a good indication of the distance between each video head. This is to ensure that a video head or "flying" erase head is not arbitrarily skipped over or accidentally chipped off during a standard cleaning procedure.

Some models of VCRs will have what is called a "flying" erase head mounted between the video heads on the upper head assembly. The flying erase head erases all previous video information from the passing video tape before new video information is recorded onto the videotape by the video heads. VCRs which contain a flying erase head are capable of high-quality editing and seamless introductions between different video scenes. It is important to use the same cleaning procedure to clean the flying erase heads as you would the video heads.

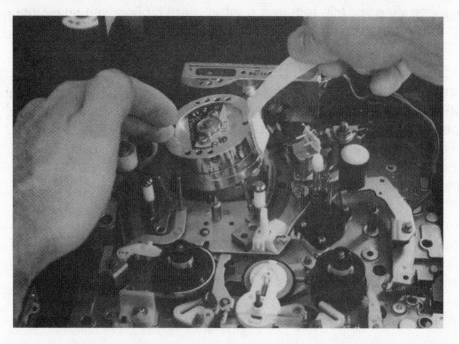

Figure 4-16. The proper way to clean video heads.

4.2.B Step Two: Cleaning the Video Heads

When cleaning the video heads on your VCR, it is important to use the proper cleaning materials. Authorized leather-tipped video head cleaning sticks can be purchased at your local audio/video supply store.

Video head cleaning sticks are bent at a slight angle as a matter of design for this particular cleaning procedure. It is important not to use a lot of pressure when applying the cleaning stick to the video heads. Also, it is highly recommended that you clean one video head at a time. This will ensure that dirt is removed properly while avoiding the chance of skipping over a potentially dirty video head.

Rotate the upper head assembly with one hand to ensure that all of video head will be cleaned. Using a leather-tipped video head cleaning stick that has been dipped in cleaning solution, hold the upper-head assembly stationary and gently apply the video head cleaning stick to the video head which you intend to clean. (*Figure 4-16.*) Gently move the video head cleaning stick in a left-to-right motion over the dirty video head. DO NOT RUB THE CLEANING STICK IN AN UP-AND-DOWN MOTION, OR YOU MAY CHIP OFF THE VIDEO HEAD!!!

Remove the cleaning stick from the side of the upper-head assembly and inspect it for dirt buildup. Chances are several distinct lines of videotape oxide are now present on the video head cleaning stick. Reapply more cleaning solution to the leather video head cleaning stick as needed.

It is recommended that you use several cleaning sticks when cleaning the video heads to ensure that the dirt being removed is not transferred to another video head.

Once the video heads have been properly cleaned, reassemble your VCR and test the playback picture quality in each of the various play speeds (SP, LP and SLP).

COMMON IDLER DRIVE
ASSEMBLIES USED IN VCRs

Chapter Five
COMMON IDLER DRIVE ASSEMBLIES USED IN VCRs

Chapter Five
COMMON IDLER DRIVE ASSEMBLIES USED IN VCRs

Located between the supply and take-up spool in the typical VCR is the idler drive assembly. (*Figure 5-1*.) The idler drive assembly plays an integral role in the VCR's overall performance. It's responsible for turning the supply and take-up spools in the proper direction, depending on the mode selected from the front panel, while the VCR is in use.

This chapter may be considered the bread-and-butter of VCR repair and cleaning. Often, each VCR you work on will need to be disassembled to this stage in order to access the idler drive assembly tire.

Believe it or not, a faulty idler drive tire is responsible for creating 75% of all faulty VCR symptoms. For this very reason, the idler drive assembly tire should always be suspected FIRST of being the cause behind the symptoms in an ailing VCR. Once you understand how to remove and access the following idler drive assemblies, you will have cornered the market in resolving the majority of the problems that affect VCRs today.

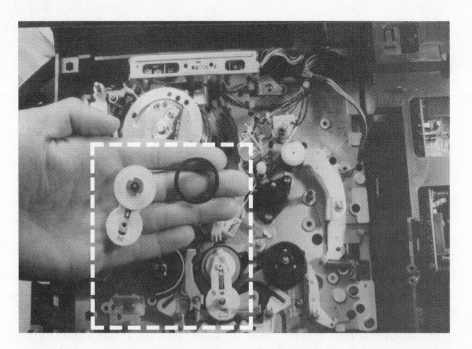

Figure 5-1. *The parts that make up the typical idler drive assembly.*

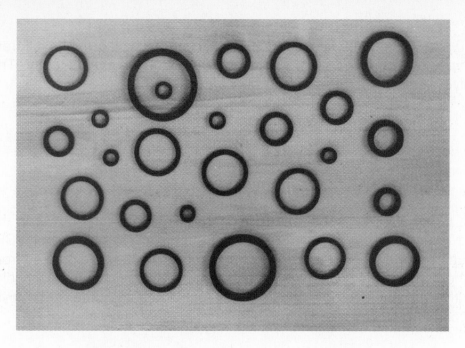

Figure 5-2. *Various types of idler drive tires.*

In most VCR formats, the idler drive assembly contains a rubber idler drive tire. (*Figure 5-2*.) The idler drive tire is responsible for providing proper traction to turn the supply and take-up spool, depending on the mode the VCR happens to be in. Through many hours of normal use, an idler tire may begin to loose traction and will begin to slip against either the take-up or supply spool, depending on the mode of the VCR. When this occurs, the VCR often exhibits a symptom. The idler drive tire can be removed and replaced with a new one.

Symptom: *The symptom for an idler drive tire that has run smooth and is slipping in place against either spool might be:*
1. *The VCR "eats" videotapes.*
2. *The VCR exhibits a sluggish Rewind or Fast-Forward function.*
3. *The VCR automatically stops playing after only a few minutes in Play mode.*
4. *The VCR may sound as if it is rewinding or fast forwarding, but the videotape is not advancing in either direction.*

If the VCR you are working on exhibits any one of the above symptoms, suspect the idler drive tire. In most cases, it will be necessary to replace the idler drive tire before the VCR will work properly in all of its various Play, Record and Review modes. A VCR may be a technical marvel, but keep in mind the integral role a simple $3.00 idler drive tire plays in the VCR's overall performance.

Typically, there are five different categories which classify the different types of idler drive assemblies used in VCRs on the market today. The five different idler drive assembly categories are:

1. Gear.
2. Pop-Out.
3. Mounted.
4. Track & Groove.
5. Imbedded.

It is very helpful to be able to recognize the variations between the different idler drive assemblies when attempting to repair your VCR. Each idler drive assembly has its own method of being extracted from the VCR for either cleaning or replacement. In this chapter, each idler drive assembly removal procedure is outlined in the order of removal complexity, not the order in which each is likely to fail first.

5.1 The Gear Drive Assembly

The gear drive assembly is denoted by a single nylon gear which interlocks with another gear located at the base of either the take-up or supply spool. (*Figure 5-3.*) Gear drive assemblies rarely fail, cause a fault, or need replacing through years of normal use. This type of drive assembly tends to be the most reliable of the five assemblies outlined here, and is often the easiest to extract from the VCR.

Figure 5-3. A gear drive assembly. (Courtesy of PRB.)

Removal Procedure for the Gear Drive Assembly

Most often, the gear drive assembly is secured on its mounting post with a small cut nylon locking washer. This nylon locking washer is removable. Simply use a pair of tweezers and lift up on one edge of the nylon washer, and it should release from its mounting post. Be certain not to lose the washer in the workings of the VCR. Once the nylon locking washer has been removed, simply lift up on the gear assembly and remove it from your VCR.

Symptom: *Gear type drive assemblies rarely fail. However, the nylon assembly of the gear has a tendency to warp or bend upward through many hours of normal use. If this is the case, the top of the drive gear will begin to rub against the metal tray of the carriage assembly. When this happens, the drive gear is inhibited and will not mesh with the gear located at the base of either the take-up or supply spool. As a result,* the playing videotape will begin to "dump" into the interior of the VCR while the VCR is in use. Also, the VCR may develop a tendency to eat tapes.

5.2 The Pop-Out Drive Assembly

The removal procedure for the pop-out assembly is very similar to that of the gear assembly. It could be said that the gear assembly is a "pop-out" assembly of sorts. However, considering that the gear drive assembly consists of a single gear and does not have a replaceable rubber idler drive tire, this gear assembly is in a category all its own. (*Figure 5-4.*)

Pop-out drive assemblies consist of a removable rubber idler drive tire which is mounted around the perimeter of a plastic wheel. In most cases, this plastic assembly wheel is secured onto its mounting post with a single cut nylon locking washer, located between the take-up and supply spools.

Figure 5-4. A pop-out drive assembly. (Courtesy of PRB.)

When a rubber idler drive tire begins to lose traction after many hours of normal use, it needs to be replaced with a new one before the VCR will work properly in its Play, Review and Record modes.

Removal Procedure for the Pop-Out Drive Assembly

The pop-out idler drive assembly is often secured to a mounting post with a cut nylon locking washer. In order to remove this locking washer, simply use a pair of tweezers and lift up on one side of the nylon washer, removing the washer from its mounting post. Be certain the washer is not lost within your VCR.

With the cut washer removed from its mounting post, lift up on the idler drive assembly in order to remove it from the VCR. The rubber drive tire, mounted around the perimeter of the assembly wheel, can then be peeled off and replaced with a new one of correct size and shape.

It is very important that a faulty idler drive tire is replaced with the VCR manufacturer's suggested replacement part number. If the replacement idler drive tire is too wide, it may jam between the two spools. If the replacement idler drive tire is too small in circumference, sufficient traction will not be made and the new tire will be heard "bouncing" between the take-up and supply spool while the VCR is in use.

Once the new idler drive tire has been mounted on the plastic assembly wheel, reinstall the assembly wheel back into the VCR. Secure the assembly wheel in place on its mounting post with the cut nylon locking washer.

It is recommended to test the performance of the VCR with its new idler drive tire installed before the VCR's lid is secured into place. This will allow you to visually inspect the performance of the new idler drive tire. You should notice an improvement in the VCR's ability to play, rewind and fast-forward if the new idler drive tire has been installed correctly.

5.3 The Mounted Drive Assembly

The mounted idler drive assembly is similar to the pop-out assembly with one exception: the mounted assembly is often secured to the chassis of the VCR with several mounting screws. (*Figure 5-5.*)

Figure 5-5. A mounted idler drive assembly. (Courtesy of PRB.)

Mounted idler drive assemblies are secured with mounting screws that are often red in color to signify their importance. Goldstar and Sharp brand VCRs often incorporate this particular drive assembly in many of their VCR formats. There are many different variations of the mounted type drive assembly on the market today; however, the removal procedure tends to be the same.

Removal Procedure for the Mounted Idler Drive Assembly

As a rule, it is very important to first release any drive belts which might be connected to the drive assembly pulley located on the lower side of the VCR. If a drive belt remains connected when you attempt to remove the idler drive assembly, damage may occur to either the VCR or to the assembly itself. Go ahead and remove any connecting lower-side drive belts at this time.

Most often, mounted drive assemblies are secured to the chassis of the VCR with two mounting screws located on the left and right side of the assembly. These particular mounting screws may be red in color to signify their importance. After disconnecting the lower-side drive belt, remove these mounting screws. The entire idler drive assembly is now free to be extracted from your VCR.

Once the idler drive assembly has been removed, the faulty idler drive tire will be accessible. Replace the faulty idler drive tire with its exact manufacturer's replacement part. To reinstall the idler drive assembly back into the VCR, simply reverse the removal steps in order and secure it into place with its mounting screws. Be certain to reconnect any lower side drive belts before testing the performance of the new idler drive tire.

5.4 The Track & Groove Drive Assembly

The track & groove assembly is a more difficult idler drive assembly to extract from a VCR. Here the plastic assembly wheel, which has the rubber idler drive tire around its perimeter, runs within a track or groove that has been cut into the chassis of the VCR. (*Figure 5-6.*) Many brands of VCR use this assembly method, and the removal procedure is typically the same regardless of the make or model.

This removal procedure tends to involve a lot of small parts, such as the take-up spool, a locking washer, a tension spring, a cassette-release bracket, and several mounting screws. It is a good idea to place the VCR parts somewhere in the order that they are removed from the VCR. This will ensure that the VCR is reassembled in the correct order once the new idler drive tire has been installed.

Figure 5-6. The plastic assembly wheel of a track & groove drive assembly. (Courtesy of PRB.)

Located between the supply and take-up spool is the plastic assembly wheel which contains the removable idler drive tire. (*Figure 5-7.*) The assembly wheel needs to be removed in order to replace the faulty idler drive tire with a new one. To create enough room for this procedure, the take-up spool will most likely need to be removed.

Begin by removing the locking washer which secures the take-up spool to its mounting post. The take-up spool can then be lifted from the VCR. Set aside the take-up spool in a safe place away from your work area, with its nylon locking washer.

Once the take-up spool has been removed, release the assembly wheel tension spring. This spring keeps the rubber idler drive tire in contact with the reel motor mounted on the underside of the VCR's chassis.

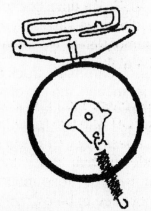

Figure 5-7. Drawing of an entire track & groove drive assembly, including the tire and wheel tension spring.

WARNING: *It is a good idea to read through this disassembly step before actually performing it. When the cassette release bracket mounting screws are removed, the drive motor (which is mounted to the lower side of the VCR chassis) will most likely "fall out". It is a good idea to hold the drive motor with a free hand when releasing these particular mounting screws. This will ensure that the motor does not swing free and break its lead wires. This also prevents the motor from damaging other components within the VCR..*

Once the assembly wheel tension spring has been released, the mounting screws that secure the cassette release bracket into place must be removed.

When a video cassette is first entered into a VCR, the carriage assembly lowers the video cassette over a brake-release bracket. The post on this bracket enters a hole located on the bottom of the video cassette shell and releases an internal "brake" within the video cassette. When this is released, the take-up and supply spools can rotate freely while the VCR is in use.

It is very important to use a free hand to catch the reel motor (which is mounted to the underside of the VCR chassis) when performing this removal procedure. The mounting screws that secure the cassette release bracket in place often secure the reel motor to the lower side of the VCR chassis. If the reel motor drops out once the mounting screws are removed, damage may occur to electrical components in the area of the reel motor.

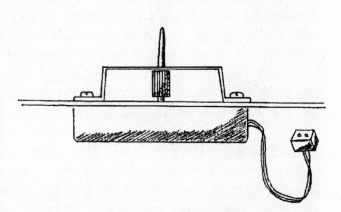

Set aside the cassette release bracket away from your work area, along with its mounting screws. If the reel motor is still secured to the underside of the chassis, go ahead and remove its mounting screws at this time.

Figure 5-8. Be sure to secure the reel motors and cassette release bracket in place with the proper mounting screws.

Once the reel motor has been removed, the assembly wheel can be extracted from the track & groove assembly. It is a good idea to use a pencil to mark the top of the assembly wheel with a "T". This will ensure that the assembly wheel is reinserted into the VCR right-side up. Remove the assembly wheel and replace the rubber idler drive tire with a new one of correct size.

To reassemble the VCR, reverse the disassembly steps. Be certain to reinstall the assembly wheel into the track & groove assembly right-side up. Using a free hand, hold the lower side reel motor in place. Secure both the reel motor and cassette release bracket in place with their proper mounting screws. (*Figure 5-8.*) Refasten the assembly wheel tension spring to the cassette release bracket and return the take-up spool to its mounting post. Secure it into place with its locking washer.

5.5 The Imbedded Drive Assembly

The fifth type of idler drive assembly is "imbedded". (*Figure 5-9*.) Imbedded drive assemblies tend to be more difficult to access than any of the previous four. In most cases, when this type of drive assembly method is used in a VCR, several obstacles must be overcome before the faulty idler drive tire can be accessed for replacement.

Figure 5-9. *An imbedded drive assembly. (Courtesy of PRB.)*

In this example, the Fisher brand VCR format is outlined. Once you have successfully performed this idler drive tire replacement procedure, other brands of VCRs containing imbedded drive assemblies will have very similar procedures.

For this particular repair procedure, the carriage assembly DOES NOT have to be removed from the VCR. Also, this procedure tends to involve a lot of small parts which should be placed aside and kept in the order that they are removed from the VCR. This will ensure that the parts are returned to the VCR in the correct order.

Figure 5-10. *The drive assembly tension spring.*

WARNING: The imbedded drive disassembly steps require the removal of two very small locking washers. It is very important that these washers do not enter the gears of the VCR or fall onto the carpet where they would certainly be lost for good. Proceed slowly and use caution when removing these washers.

Figure 5-11*. The tension spring and take-up spool have been removed.*

Begin the removal procedure by first releasing the drive assembly tension spring which connects the assembly wheel to the chassis of the VCR. (*Figure 5-10.*) This particular spring keeps proper tension on the idler drive tire that is imbedded beneath the VCR's chassis. After removing the idler drive tension spring, remove the tension spring located to the upper left of the take-up spool. Set aside the two springs in a safe place away from your work area.

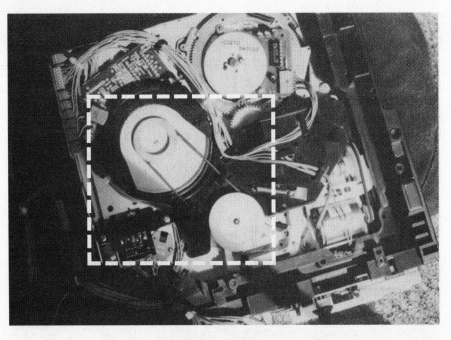

Figure 5-12*. The connecting drive belt between the capstan motor and the drive assembly pulley.*

Next, the take-up spool needs to be removed. This will provide enough room for the idler drive assembly to be maneuvered out of the VCR. The take-up spool can be extracted without having to remove the VCR's carriage assembly. Simply unfasten the nylon locking washer which secures the take-up spool to its mounting post, and the take-up spool can be lifted out of the VCR. Believe it or not, these are the only parts removal steps from the top of the VCR. (*Figure 5-11.*)

With the VCR on its back (top) and its lower pan removed, continue with the disassembly steps by removing the large drive belt that connects the capstan motor to the drive assembly pulley. (*Figure 5-12.*) Once this drive belt has been removed, disconnect the large tension spring that is secured to the transfer assembly arm.

The next disassembly step requires removal of the plastic cover guard in order to provide more room to access the faulty idler drive tire. Not all VCRs will have such a cover guard. Located beneath the various connecting wires, hidden from sight, is the cover guard mounting screw that needs to be removed. Go ahead and remove this screw. (*Figure 5-13.*) The cover guard is now free to be positioned out of the way.

With the cover guard out of your way, the next step is to remove the nylon locking washer that secures the VCR's assembly gear to its mounting post. (*Figure 5-14.*) After removing this locking washer, the assembly gear is free to be lifted from the VCR. This is the transfer

WARNING: When reinstalling the idler drive tire assembly, it is very important to correctly seat the hub of the assembly within its groove cut into the chassis before the VCR is reassembled. If this particular hub is not seated correctly and the performance of the VCR is tested, chances are the VCR will exhibit a symptom. The symptom might be sluggish Rewind and Fast-Forward modes, or the videotape may not advance.

Figure 5-13. *Removing the cover guard mounting screw.*

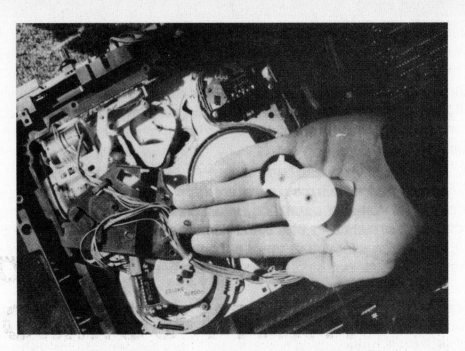

Figure 5-14. *The assembly gear locking washer.*

assembly that provides proper traction to the idler drive tire while the VCR is in use. Be certain to clean the nylon traction wheel located at the end of this particular assembly. Set aside this transfer gear with its nylon locking washer in a safe place away from your work area.

The second nylon locking washer to be removed secures the assembly that contains the removable rubber idler drive tire. After removing this nylon locking washer, the idler drive assembly will be free to be lifted from the VCR. Be certain the hub of the assembly is maneuvered out of its groove before lifting up.

With the drive assembly removed, simply peel the old rubber idler drive tire off and replace it with a new one. Be certain the new idler drive tire is of correct size before reassembling the VCR. This particular idler drive tire is thin, almost the same dimension as a square loading belt.

Once the new idler drive tire has been installed, simply return the various parts to the VCR in the reverse order of which they were removed. Be certain to correctly seat the idler drive assembly into its groove, and replace the locking washers, tension springs, plastic cover guard, drive belt and take-up spool.

Once the VCR has been reassembled, test its performance with the new idler drive tire installed. The VCR should continue to run well for many hours of normal use.

Chapter Six

THIRTY COMMON VCR PARTS & THEIR SYMPTOMS

Chapter Six
THIRTY COMMON VCR PARTS & THEIR SYMPTOMS

It may seem as though there are nearly a thousand different variables to take into consideration when troubleshooting the typical VCR for the reasons behind its faults. However, this is often not the case. When a VCR does exhibit a symptom, chances are the faulty part can be isolated after considering only three or four different areas within the VCR.

This chapter outlines thirty common parts that represent nearly 85% of all VCR faults. Believe it or not, an experienced VCR technician can often isolate the root of a symptom simply from the description provided by a calling customer, as outlined in Chapter Ten.

The real trick to VCR repair and cleaning is to understand the role that each individual part plays in the VCR, and to be able to identify the symptom for that particular part when it is dirty, requires adjusting, or has failed.

The term "symptom similarity" describes two or more different VCR parts that exhibit the same symptom. For example, the symptom for a slipping idler drive tire may resemble the symptom for a failed take-up spool sensor. That is to say, the VCR will automatically default to the Stop mode a few seconds after the Play command has been selected. Although symptom similarity may exist between these two different VCR parts, their removal procedures and the cost of replacing them vary considerably. This is a very important aspect to take into consideration when troubleshooting your VCR. If you do not recognize the root of the original symptom, and arbitrarily begin to replace parts hoping this will correct the problem, you will soon deplete your supply of good parts, customers, and money.

If you are unsure as to what is causing the symptom in your VCR, use this chapter as a part-reference guide, and call a VCR repair center in your area. The VCR technician should be able to shed some light on your problem if you inform the technician of how far you have progressed in the disassembly process. Be certain to write down the technician's answer just in case you run across this symptom again in the future. If you have any questions, now is the time to ask. Inform the technician of your theories behind the symptom. Novices to VCR repair and cleaning automatically assume the VCR's microprocessor has failed and is creating the original symptom; however, the majority of all VCR faults rarely originate from the VCR's microprocessor. Be aware that many VCR repair centers have been plagued by managers who take advantage of the general public's naivete and are quick to exaggerate a symptom in order to inflate an otherwise simple repair.

Now, without bashing the honest repair centers, let's consider the thirty common parts within the typical VCR and their symptoms:

6.1 Harness & Ribbon Cables

Harness and ribbon cables are used extensively throughout the typical VCR in order to transfer vital electrical information to the various parts of the VCR.

Harness cables are individual wires with a "female" connector secured at both ends. The female connector fits onto a male connector that is permanently mounted to a PC board in the VCR. A common location for the female/male harness cable connector is at the rear corner of the carriage assembly. Of the different female/male connectors used within the typical VCR, the connector located at the corner of the carriage assembly is often left unconnected as an oversight after a standard cleaning or repair procedure, resulting in a fault.

Ribbon cables, on the other hand, are a direct result of the computer age. Ribbon cables consist of individual wires which are sealed side-by-side in a protective jacket. Often, the bare lead wires of the ribbon cable simply insert into a connector that is soldered into place on the VCR's PC board. The end of the ribbon cable, where the lead wires are inserted into their connector, is a common area to search for an intermittent fault.

Symptom: *The symptoms for ribbon cable lead wires which may have worked themselves out of their connector might be:*
1. *The VCR may randomly stop without notice while in the Play or Record mode.*
2. *The VCR may refuse to go into a particular mode selected from the front panel.*
3. *The VCR will not receive or eject a video cassette.*

To remedy this problem, aside from reinserting the bare ribbon cable lead wires into their connector, it may be necessary to remove the connector and hard-solder each individual lead wire to the PC board on which the connector is mounted. This will eliminate the loose ribbon cable problem and reduce the chance of an intermittent symptom occurring again in the future.

6.2 Power Supply Fuse

All VCRs on the market today use at least one fuse in their power supply as a safety feature to protect against power surges through the VCR's power cord. (*Figure 6-1.*) The power supply fuse is designed to fail if the incoming electrical current exceeds the specified amount for the electrical components of the VCR. The typical VCR fuse ranges between 1.2 and 2.5 amps/125 volts. It is very important to replace a power supply fuse with one of exact or lesser amperage. This is to ensure that the VCR's electrical components are not damaged once the power supply cord is inserted into a wall outlet and the VCR is powered on.

Symptom: *The typical symptoms of a failed power supply fuse are:*
1. *The VCR has no power.*
2. *A specific function of the VCR does not respond, such as a drive or loading motor.*
3. *The VCR may intermittently power-up if the clip that holds the power supply fuse is loose.*

Keep in mind that a VCR may incorporate five or six different fuses that are designed to protect individual PC boards within the VCR. If one of these fuses fails, the VCR may power-up, but a particular area in the VCR (such as the clock display, a loading motor, or a drive motor) may be dormant. The VCR will not work properly again until the failed fuse is replaced with a new one of correct size and amperage.

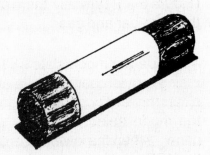

Figure 6-1. *Drawing of a typical power supply fuse.*

6.3 Worm Gears

Worm gears play an integral role in the overall performance of the VCR. (*Figure 6-2.*) Nearly every VCR format will have a worm gear among its other gears in order to accomplish a specific mode selected from the VCR's front panel. Typically, worm gears tend to be located on the lower side of the VCR's chassis. A traveler arm is used in conjunction with the worm gear. The traveler arm has a small post protruding from it, and runs within the groove of the worm gear. After many hours of normal use, the wall of the worm gear groove may become soft and "collapse". Also, hardened grease may accumulate in the groove of the gear over time. When this occurs, the post of the traveler arm may either bind-up or jump out of its track on the worm gear during the execution of a particular front panel command, resulting in a fault. It is a good idea to clean the groove within the worm gear and regrease it before assuming that the gear is faulty and needs replacing.

Symptom: *The symptoms for a worm gear groove which is either obstructed with a foreign object or whose inner wall has collapsed are:*

1. *The VCR may attempt to execute a particular front panel command, but the assembly gears make a loud snapping or grinding sound. The VCR does not go into the selected mode.*
2. *The VCR may power-off after attempting a particular front panel command.*
3. *If the VCR guide posts "die in the out position" during Play or Record mode, always suspect either an obstructed or collapsed worm gear that is having difficulty rotating.*

Figure 6-2. *A worm gear. (Courtesy of PRB.)*

6.4 Mode/State Switch

Mode/state switches are common to every VCR on the market today. (*Figure 6-3*.) Typically, there are two types of VCR mode/state switches; the slide-bar and gear.

The slide-bar mode/state switch will have a single nylon bar which is maneuvered either forward or backward depending on the mode selected on the VCR's front panel. When the lower assembly gears of the VCR rotate, the nylon bar of the mode/state switch is mechanically positioned into place so the VCR can execute the particular mode selected from the front panel.

Figure 6-3. *Illustration of a common style of mode state switch.*

The gear mode/state switch has a nylon gear instead of a slide bar. This gear is directly connected to the assembly gears of the VCR. When a particular mode is selected from the VCR's front panel, the assembly gears rotate which, in turn, rotate the gear of the mode/state switch, thus putting the VCR into the desired mode.

Symptom: *The symptoms of a mode/state switch that is either dirty or has failed might be:*

1. *The VCR may seem "confused," or may hesitate before attempting a particular command selected from the front panel.*

2. The VCR may default to an unspecified mode such as Stop,
   Play or Rewind without prior warning.
3. The VCR may receive a video cassette but will not go into any
   mode selected from the front panel.

Many VCR mode/state switches consist of two nylon halves that clip together and can be disassembled as a matter of design. When the two halves are separated, the inner contacts and sweep arms can be accessed in order to remove dirt which has accumulated through many hours of normal use. It is recommended that the inner contacts of the mode/state switch be cleaned thoroughly before assuming that the VCR's mode/state switch is indeed faulty and needs to be replaced.

6.5 Square Belts

Very rarely will a VCR be missing a drive or loading belt. Typically, there are four circumference categories for square belts that are used in VCRs on the market today. The four categories are:
1. Very Small.
2. Small.
3. Medium.
4. Large.

Let's consider the location and use of each particular belt, since the symptoms of a slipping drive or loading belt is typically the same regardless of its circumference.

6.5.A Very Small Square Belts
The square VCR belt with a very small circumference is most often used on the carriage assembly as a loading belt. This particular belt should be cleaned with cleaning solution on a regular basis to ensure that the VCR continues to receive and eject video cassettes properly. Call it planned obsolescence, call it bad karma; at any rate, the small VCR belts tend to perform the most work and thus are most likely to wear out.

Another location within the VCR for very small belts might be on the lower side of the VCR chassis. When the VCR is placed into either the play or record mode, this size belt is often used in order to mechanically extend the guide posts forward into their full V-block position.

6.5.B Small Square Belts
The next circumference of drive or loading belt is the small square belt. This particular size belt is often located around the take-up spool as a counter belt, or on the lower side of the VCR's chassis as a drive belt.

A good example of this can be found in Panasonic and Fisher brand VCR formats. Often, a pair of tweezers is all that is required to remove this drive belt from its drive motor and connecting pulley.

6.5.C Medium Square Belts

The third common size square belt is much larger in circumference than the previous two, and is often used to drive the idler drive assembly or the capstan post via the capstan motor. This particular size drive belt tends to be located on the lower side of the VCR chassis. It is a good idea to inspect this size belt for wear and tear since it provides proper traction to rotate the idler drive assembly. After many hours of normal use, this drive belt may run smooth, break in half, or fall off completely, resulting in a fault.

6.5.D Large Square Belts

The large square belts tend to be "workhorses," and are in continuous use with each command selected from the front panel of the VCR. Earlier "piano-key" VCR formats often incorporated several belts of this circumference. This size drive belt is susceptible to stretching and deteriorating through many hours of normal use. When this occurs, a belt of this size may slip off its pulley and wrap around the various parts which make up the VCR. The VCR is rendered useless until the belt is replaced with a new one of correct circumference.

Symptom: The symptom for a failed loading or drive belt might be:
1. The VCR refuses to receive or eject a video cassette.
2. A loud "squelching" noise emits from the VCR when it is powered on.
3. The VCR is eating videotapes.
4. The VCR exhibits a sluggish Rewind or Fast-Forward mode, or the VCR seems to go into mode but powers off immediately afterward.
5. A still picture or "freeze frame" is present on the television screen when a video cassette is playing, and the VCR soon powers off.

To remedy any of these symptoms, be certain to replace faulty drive or loading belts with new ones that are 5% smaller in circumference than belts being replaced. However, it is always best to use the manufacturer's part number when reordering a single belt or belt kit for your VCR. By using manufacturer part numbers when reordering belts and belt kits, you avoid second-guessing your original repair work, and reduce the chance of another slipping-belt symptom in the near future.

Figure 6-4. Various types of idler drive tires and assemblies.

6.6   Idler Drive Tires

The VCR's rubber idler drive tire plays an integral role in the VCR's overall performance. (*Figure 6-4.*) Think of the idler drive tire as a tire on your automobile. The idler drive tire provides traction to rotate either the take-up or supply spool. Through many hours of normal use, this particular part in the VCR tends to run smooth and eventually lose its ability to rotate either spool. When this occurs, the VCR will most likely exhibit a fault.

For a complete removal procedure and further information about the VCR's idler drive tire, it is a good idea to read Chapters Five and Seven.

Symptom: *The symptom for a failed idler drive tire might be:*
  1. *The VCR eats videotapes.*
  2. *The VCR exhibits a sluggish Rewind or Fast-Forward mode.*
  3. *The VCR may sound as if it is rewinding a video cassette, but the videotape is not advancing.*
  4. *The VCR stops or powers off a few minutes after being placed in the Play, Rewind, Fast-Forward or Record mode.*
  5. *The VCR may refuse to either accept or eject a video cassette.*

To remedy any one of these symptoms, the idler drive tire will need to be replaced with a new one. It may be possible to clean the perimeter of an idler drive tire with cleaning solution as a "quick fix" in order to get the idler drive tire to continue providing traction for the duration of a rental movie, for example. However, this is generally not recommended since the slipping idler drive tire may begin to eat the videotape without prior warning.

Believe it or not, a failed idler drive tire accounts for nearly 75% of all VCR symptoms. Although the VCR seems to be an electro-mechanical marvel, always keep in mind the integral role a $3.00 rubber idler drive tire plays in its overall performance.

6.7 Pinch Roller

As with every VCR part, the pinch roller plays an integral role in the performance of the VCR. All VCR formats incorporate a pinch roller within their tape path. (*Figure 6-5.*) Pinch rollers are made of rubber and tend to become smooth and "pitted" after many hours of normal use if they are not cleaned on a regular basis.

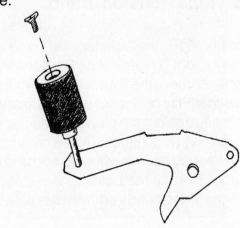

Figure 6-5. *Drawing of a pinch roller, attached to its assembly arm by a single mounting screw.*

As a matter of design, the pinch roller is mechanically positioned into place tightly against the capstan post when either the Play or Record command has been selected from the VCR's front panel. As the capstan post rotates, the combination of both capstan post and pinch roller pull the videotape through the VCR's tape path.

The pinch roller is often secured to its assembly arm with either a single mounting screw or nylon cap inserted through the center of the pinch roller. Both the mounting screw and nylon cap are removable so that the pinch roller can be replaced if it is faulty. This area of the VCR's tape path tends to act as a magnet for dirt, rug fibers, videotape oxide, and oil. The pinch roller may require 10-15 cleaning swabs to sufficiently remove this dirt buildup if the pinch roller is not cleaned on a regular basis.

Symptom: *The symptom for a faulty pinch roller might be:*
1. *The playback video picture tends to come and go with intermittent patches of "snow".*
2. *If an upper or lower edge of the pinch roller becomes warped or pitted, both the audio and video portion of a playing video tape may become distorted.*
3. *An upper or lower edge of the passing videotape may become scalloped or wrinkled as it passes between the pinch roller and capstan post.*

If any one of these symptoms occur, the pinch roller may no longer be flush against the capstan post and the videotape is most likely being pulled through the VCR's tape path at a slight angle, resulting in a fault.

To remedy these symptoms, the pinch roller either needs to be cleaned thoroughly or replaced with a new one of correct width and circumference.

6.8 Tape Tension Band

In every VCR, the tape tension band is located around the base of the supply spool. (*Figure 6-6.*) This tension band was set to manufacturer specifications at the time of the VCR's assembly. When the VCR is placed into either Play or Record mode, a tension arm is mechanically positioned into place against the passing videotape. As a matter of design, the tape tension band is connected to the base of this tension arm. The tape tension band keeps proper tension on both the supply spool and the tension arm in order to keep the passing videotape from "sagging" or drifting away from the video heads and various tape path parts.

Through many hours of normal use, the glue which secures the felt pad located on the inside of the metal tension band may dry out, and the felt pad may fall off, resulting in a symptom.

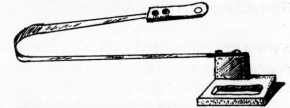

Figure 6-6. Drawing of a common tape tension band, which is located around the base of the supply spool.

Symptom: *The symptom for a tape tension band problem might be:*
1. *The video playback picture may flutter, randomly jump, or have horizontal lines running through it. Adjusting the tracking control knob does not correct the problem.*
2. *The playback picture of a recent recording may exhibit intermittent patches or bands of snow.*

To remedy these problems, the tape tension band will need to be replaced with a new one. Before removing a faulty tape tension band from the VCR, it is a good idea to outline the perimeter of the tension band with a pencil while it is still secured to the chassis of the VCR. Once the faulty tape tension band has been removed, this outline can be used as a reference for correctly seating the new tape tension band back into position.

If the tape tension band can be salvaged, it is a good idea to resecure the felt pad with bonding glue or cement as a last resort if it has freed itself from its metal band. Often this is all that is required to correct a fault if the tension band is not broken.

6.9 RF Modulator 3/4 Channel Switch

Located on the back of the VCR are two protruding posts and a 3/4 channel switch. (*Figure 6-7.*) Both are part of the VCR's radio frequency modulator, or RF modulator. The RF modulator receives "unreadable" video information through the "Antenna In" post, deciphers this signal into "readable" video information, and makes the signal available to either the Channel 3 or 4 setting on your television set via the "Out To TV" post on the RF modulator.

The 3/4 channel switch on the RF modulator should always correspond to the channel setting on your television in order to view a clear television broadcast signal through your VCR. The majority of VCR owner manuals suggest that Channel 3 be allocated as the channel setting for both the RF modulator and the television set.

Symptom: *If the 3/4 channel setting on the RF modulator does not correspond with the selected viewing channel on the television set, the symptom for this might be:*
  1. *The playback picture from the VCR to the television is wavy and distorted.*
  2. *Any broadcast station recorded through the VCR would most likely play back distorted or as an entire field of snow.*

Believe it or not, something as simple as the VCR's 3/4 channel switch being accidentally set on the wrong channel setting could result in hours of misguided "tweaking." Be certain the RF modulator 3/4 channel setting

Figure 6-7. Several different RF modulator 3/4 channel switches.

corresponds with the television viewing channel, either 3 or 4, to remove any doubt that the RF modulator is indeed faulty.

6.10 Static Guard

Located on top of the upper head assembly or under the lower drum motor is a static guard. When a VCR plays a videotape, static electricity tends to be generated: a direct result of the passing videotape coming in contact with the various tape path parts. Since the video heads are very sensitive to any form of static electricity, the static guard is designed to defuse static buildup in an attempt to protect the video heads from being damaged. If the static guard is bent, either as a direct result of replacing the video heads or during the cleaning process of the tape path parts, a fault may occur.

Symptom: *The symptom for a bent static guard might be:*
1. *Snow "bullets" or "meteors" randomly shooting across the television screen while the VCR is playing a videotape.*
2. *If the static guard is bent or vibrating in place against the upper head assembly, the VCR may exhibit a loud "wringing" or scraping sound when the upper head assembly rotates at its proper Record or Play speed.*

To remedy this problem, be certain the static guard is not bent in any way. Also, make a point to remove any hardened grease and accumulated dirt from beneath the static guard where it comes in contact with either the upper head assembly or the lower drum motor.

6.11 Timer Switch

The timer switch is included in this section of common VCR parts since it is a potential hot spot for faults. (*Figure 6-8.*) Whenever a VCR is programmed to record a television show at a future time and date, the timer switch is set to the "Timer On" position. Only until the timer switch is set to this position will the VCR record automatically. If the timer switch is unintentionally positioned to the "Timer On" setting when the front panel of the VCR is reinstalled after a repair or cleaning, a fault may occur.

Symptom: *The symptom might be:*
1. *The VCR may not accept a video cassette.*
2. *The VCR may eject any video cassette missing a record-safety tab, such as rental tapes.*
3. *The VCR may not power-up for normal Play or Record use until the timer switch is in the "Timer Off" position.*

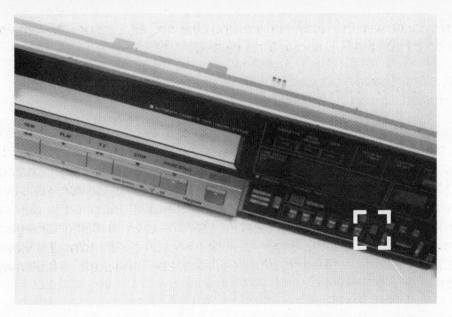

Figure 6-8. Location of a timer switch of a VCR's front panel.

It is always a good idea to inspect this area of the VCR when reinstalling the VCR's front panel. Test the movement of the timer switch to ensure that is not hindered in any way.

6.12 VCR Tension Springs

Nearly every VCR on the market today uses one or more tension springs to keep proper tension on a specific part within the VCR. (*Figure 6-9.*) There are many uses for tension springs throughout the typical VCR.

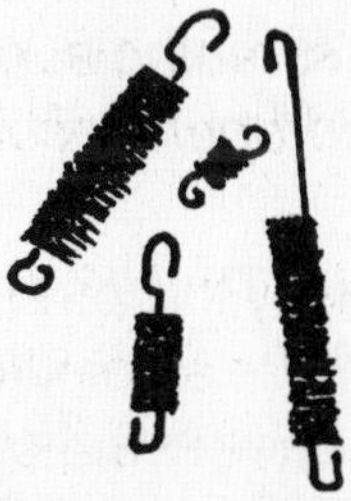

Figure 6-9. Drawing of tensions springs of various sizes.

Tension springs are often used to keep the idler drive tire in contact with its drive motor. Tension springs are often located at the base of the tension arm, on the lower side of the VCR chassis securing a slide arm into place, on one of the break arms located on either side of the supply and take-up spools. If a tension spring is left unfastened after the removal procedure of a faulty part, and the performance of the VCR is tested, a fault may occur.

Symptom: The symptoms for an unhooked tension spring might be:
1. *The VCR may attempt to execute a mode but the videotape is not advancing through the tape path.*
2. *The audio and/or video portion to a playing videotape may sound and look as if it is being played underwater.*
3. *The VCR may begin to eat videotapes.*

To remedy any one of these problems, be certain all of the tension springs have been returned to their original location before testing the performance of your VCR. Also, it is NOT a good idea to spray the inside of the VCR with compressed air once the VCR lid or lower pan has been removed. If a tension spring has come unfastened and is creating the original symptom, it does not make sense to spray it out of the VCR and onto the floor where it would certainly be lost for good.

6.13 Various VCR Sensors

Manufacturers of VCRs often incorporate various sensors throughout the VCR. Typically there will be a sensor located under the take-up spool, as well as a start-and-end sensor mounted on either side of the carriage assembly. VCR sensors often come in a variety of shapes and sizes. It is very important to replace a failed sensor with its exact manufacturer's replacement part. Although VCR sensors may look similar in both size and shape, they are often not interchangeable.

When a VCR fails, the VCR's microprocessor notes this and often defaults to a "no-play" mode as a safety feature to protect the videotape from being damaged by the VCR. Only until the failed sensor is replaced will the VCR work properly in all of its various Play and Record modes.

Symptom: *The symptom for either a failed start or end sensor, located on the VCR's carriage assembly might be:*
  1. *The capstan motor and supply spool immediately begin rotating once the VCR is powered on from the front panel.*
  2. *The VCR may not accept a video cassette and may power-off after several seconds.*

The symptom for a failed take-up spool sensor looks very similar to the symptom for a slipping idler drive tire. It can be said that "symptom similarity" exists between these two different VCR parts.

Symptom: *The symptom of a faulty take-up sensor might be:*
  1. *The VCR may begin to advance the videotape for several seconds while in the Play or Record mode, but powers-off shortly afterward.*
  2. *The digital tape counter does not increment before the VCR powers-off.*

To save valuable troubleshooting time, it is often possible to rotate the take-up spool to detect whether or not the digital counter is incrementing. If the digital counter is NOT incrementing, chances are the take-up spool sensor is faulty and will need to be replaced before the VCR will work again properly.

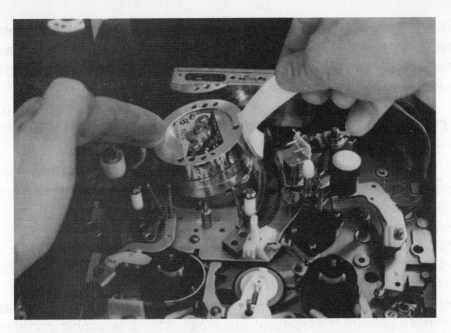

Figure 6-10. *The proper way to clean video heads.*

6.14 Video Heads

Consider the video heads to be the "eyes" of your VCR. Without video heads, the VCR would be unable to either read from or record to the passing videotape. The video heads should NEVER be touched directly with your fingers. Oil deposits from your fingertips can severely affect the performance of the sensitive video heads. After many hours of normal use, the video heads may accumulate dirt as a result of coming in contact with the passing videotape. The video heads should always remain dirt and oil free for proper playback and recording clarity. (*Figure 6-10*.)

Also, NEVER attempt to clean the video heads with a cotton or foam-tipped cleaning swab. The cotton fibers and porous foam tip may snag one of the video heads and can easily chip it off. Leather-tipped video head cleaning sticks are bent at a slight angle as a matter of design and should be used exclusively to clean the video heads.

Symptom: *When the video heads accumulate videotape deposits, the fault is often very obvious to detect. The symptom for dirty video heads might be:*
   1. *Snow or black-and-white "fuzz" running horizontally across the television screen.*
   2. *The video portion of a playing videotape cannot be seen, but the audio portion can be heard.*

Since the video heads operate independently from the audio heads, the audio portion of a playing videotape can often be heard while the video portion is snowed out.

If the video heads still exhibit a snowy picture after several cleaning attempts, chances are the video heads are faulty. It is a good idea to reference section 4.2 of Chapter Four before assuming that the video heads are indeed faulty and need to be replaced.

6.15 Nylon Locking Washers, E and C Clips

VCR manufacturers use several different methods for securing the various parts of the tape path to the chassis of the VCR. There are typically three different methods:
1. Nylon locking washers.
2. E clips.
3. C clips.

Often, limiter arms, pinch rollers, assembly gears and drive pulleys are secured with one of these particular methods. It is very important to resecure any VCR part that may have been removed during a standard cleaning or repair procedure. If something as simple as a nylon locking washer or C clip is left out of the VCR, and the performance of the VCR is tested, a fault may occur.

Symptom: The symptom for a missing nylon locking washer, E clip or C clip might be:
1. *A loud rattle may be heard while the VCR is in use if a take-up or supply spool locking washer is missing.*
2. *The VCR may attempt to play a videotape but powers-off if an integral C clip is left off a lower side gear.*
3. *The VCR may begin to wrinkle an edge of the playing video-tape if the tape tension arm is not properly secured to the chassis of the VCR with its E clip.*

6.16 Audio Control Erase Head Assembly

Located on the right side of the typical VCR tape path is a dual-head assembly that is mounted to an adjustable plate. This adjustable plate is secured to the metal chassis of the VCR. The dual-head assembly contains the audio control erase head, or ACE head.

It is very important that the audio control erase head is kept both dirt and oil free to ensure proper record and playback clarity. Since the passing videotape comes in contact with the ACE head at all times while the VCR is in use, dust, pet hairs, dirt deposits and even video cassette labels have a tendency to accumulate on the leading edge of this particular head assembly.

Symptom: The symptom of an audio control erase head which has accumulated dirt through many hours of normal use might be:
1. The audio portion of a playing videotape may be very faint.
2. The VCR may have difficulty reading the encoded information on the lower edge of the passing videotape, and may randomly alternate between the various play speeds (SP, LP and SLP).
3. The playing videotape may exhibit an alien or "underwater" sound quality.

To remedy these problems, use a cotton-tipped swab dipped in cleaning solution to remove accumulated dirt from this location within the VCR's tape path. Be certain to clean under the slight overhang which protrudes from the top of this assembly. The audio head is located under this overhang and is often overlooked by the novice VCR repair and cleaning person.

6.17　Guideposts

The VCR's guideposts play an integral role in the overall picture quality of a playing videotape. Guideposts are used in all VCR formats to extract the videotape from the plastic cassette shell and apply the videotape to the VCR's tape path.

Located at the base of each guidepost is an Allen nut which must be loosened with a VCR Allen wrench before the adjustment nut (located on the top of each guide post) can be rotated. If the base-locking Allen nut is not loosened before an adjustment procedure, the guidepost may become stripped and may not remain stationary while the VCR is in use. However, through many hours of normal use, the base-locking Allen nut on either guidepost may loosen on its own, which will result in a fault.

Symptom: The symptom for a tape path guidepost that is out of alignment might be:
1. The upper or lower edge of the playback picture is wavy or has bands of lines running horizontally across it.
2. The audio portion of a playing videotape may be very faint.
3. Bands of snow may intermittently appear on the playback picture when playing a recently recorded videotape.

Often, when the left guidepost is out of alignment with the rest of the VCR's tape path, bands or horizontal lines begin to appear in the UPPER portion of the playback picture. On the other hand, when the right guidepost is out of alignment with the VCR's tape path, bands or horizontal lines may begin to appear in the LOWER portion of the playback picture.

WARNING: For a detailed explanation of the ACE head assembly, it is a good idea to review section 4.1.G. of Chapter Four before attempting to remedy a potential fault which you suspect is originating from this area. Novices to VCR repair and cleaning tend to "tweak" the base adjustment screws of the ACE head hoping this will correct a symptom. Often it does not, and the VCR will require help from a professional service center.

WARNING: It is very important to proceed very slowly when adjusting this area of the VCR tape path. Be certain to read the procedure for adjusting the VCR's guideposts several times before attempting this adjustment procedure on a working VCR. The slightest random "tweaking" to either guidepost may have a severe effect on both the audio and video portion of a playing videotape.

6.18 Drive & Loading Motors

VCRs on the market today will have a drive or loading motor in order to accomplish a specific command. (*Figure 6-11.*) In most cases, drive & loading motors are directly connected to assembly gears via a rubber drive or loading belt. Through many hours of normal use, drive & loading motors tend to overheat, and eventually freeze-up or stop rotating. The average lifetime for a loading motor is about six to eight years.

Figure 6-11. *Drawing of a typical drive/ loading motor.*

Symptom: *The symptom for a drive or loading motor that has failed might be:*
1. *The drive or loading motor is very hot to the touch.*
2. *A power supply fuse may continue to fail when the VCR is in use.*
3. *The VCR will not receive or eject a videocassette.*
4. *The VCR may attempt a command selected from the front panel but immediately powers off.*
5. *The shaft of the drive or loading motor is frozen or does not rotate.*

In either case, a faulty loading or drive motor must be replaced with a new one before the VCR will work properly again.

6.19 Various VCR Switches

As a matter of design, the various switches found in the typical VCR tend to be located toward the front where the video cassette enters and is seated in the VCR for playing or recording.

Housed within the carriage assembly are a host of command switches which initiate the loading and unloading process. When a video cassette is entered into the VCR, often a loading switch will be activated. When the Eject command is selected from the VCR's front panel, a different switch may be used in order to initiate the unloading process. VCRs also contain a "cassette-in" switch that notifies the microprocessor that a video cassette is seated within the VCR. If a particular switch is either dirty or faulty, the proper signal will not be sent to the microprocessor and a symptom will occur.

Symptom: *The symptom of a failed or dirty loading switch might be:*
1. *The carriage assembly will not receive a video cassette.*

2. The carriage assembly may quickly eject a video cassette once it is seated within the VCR.
3. The VCR may not go into a particular mode if the "cassette-in" switch is not functioning properly.

It is a good idea to inspect each switch on the VCR's carriage assembly to ensure that a particular contact has not accumulated excessive dirt or is bent.

6.20 Flat Drive Belts

Aside from the typical various size square drive and loading belts found within the VCR, flat belts are used to accomplish particular tasks selected from the VCR's front panel. Flat VCR drive belts are often much wider and thinner than square VCR drive and loading belts.

Symptom: It is very important to use the correct manufacturer's part when reordering flat drive belts for your VCR. If a too-tight flat drive belt is installed, the drive motor that it is connected to may overheat. Also, a power supply fuse may continue to fail until that particular belt is replaced with one of correct size. Also, if a too-wide flat drive belt is installed beneath other belts, a fault may occur. The symptom for this might be that the gears or pulleys connected to the other belts may hesitate or jitter while the VCR is in use. This is due to the fact that the crossover belts are making contact with the newly-installed flat belt.

Other symptoms might include:
1. *Garbled audio and/or video on a playing videotape.*
2. *Patches of snow while the videotape is in the Play mode.*
3. *The VCR sounds as if it is accomplishing a particular mode, but the videotape is not advancing through the tape path, and the VCR soon powers off.*

6.21 STK IC Voltage Regulator

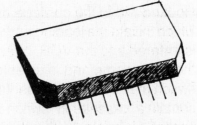

STK integrated circuits, or STK ICs, are used within the VCR power supply to regulate the various voltage levels used throughout the VCR. (*Figure 6-12.*) This particular function makes them very susceptible to failure.

Figure 6-12. Drawing of an STK voltage regulator IC.

Manufacturers of VCRs apply a liberal coat of heat-sink grease to the backing of each STK IC. Since STKs are considered to be hot-running ICs, this coating of heat-sink grease acts as a buffer to dissipate heat buildup and prolong the STK IC's useful life. However, after many hours of normal use, the STK IC may overheat, resulting in a symptom.

Symptom: The symptom for a failed STK IC might be:
1. The upper head assembly does not rotate when the Play command is selected.
2. If your VCR does have an STK IC, the loading motors can often be heard initializing once the VCR power supply cord is plugged into a wall outlet. If the motors are NOT heard initializing, the STK IC might be faulty.
3. The power supply light may blink for seven seconds and then stop. The VCR does not power-up at all.

To ensure that the new STK IC is installed correctly and has a liberal coat of heat-sink grease applied to its backing, read the replacement procedure for a faulty STK IC as outlined in Chapter Seven to ensure that an important step is not overlooked.

6.22 Dew Sensor

Many early brands of VCRs have a dew sensor built into their electrical format as a safety precaution to warn against hazardous levels of moisture buildup. (*Figure 6-13*.) If the room humidity exceeds a specified level, moisture may condense on the VCR's video head assembly and other internal tape path parts.

Figure 6-13. *Drawing of a VCR dew sensor.*

Symptom: If a playing videotape were to come in contact with moisture buildup, either on the video head assembly or other tape path parts, the passing videotape would quickly stop advancing. With the upper head assembly rotating at 1800 rpm, the result can be a tangled knot of videotape that has literally wrapped itself around the rotating upper head assembly.

Be certain not to touch the surface of the dew sensor with your fingers. Oil from your fingertips may effect the performance of the dew sensor which could result in the VCR eating a videotape.

It is NOT recommended to use a low-set hair dryer to dry the dew sensor if the dew indicator light remains lit. Also, do not disconnect the dew sensor in order to bypass a potential fault. The dew sensor has been installed for a good reason.

WARNING: *If sewing machine oil or VCR grease were to come in contact with the surface of the dew sensor, the dew indicator light (located on the front panel of the VCR) would most likely remain lit and the VCR would refuse to go into a particular mode.*

6.23 Full-Erase Head

Located to the left of the VCR tape path is the full-erase head. (*Figure 6-14.*) The full-erase head erases all previous video information from the passing videotape when the VCR is in its Record mode. Video cassette players or VCPs do not contain such erase heads since they are not equipped to record information onto videotape.

Figure 6-14. *Drawing of a tape path full-erase head.*

It is very important for the surface of the full-erase head to remain dirt and oil free for maximum erasing capabilities. Like the audio control head, the leading edge of the full-erase head tends to be a magnet for dust, pet hairs, videotape oxide, and even video cassette labels. If this particular head is not cleaned on a regular basis, the accumulated dirt may have an adverse effect on the recording quality of your VCR.

Symptom: *The symptom for a dirty or failing full erase head might be that previous audio and/or video information is still present on a recent recording.*

What is happening here is that the electrical signal responsible for erasing the previous video information from the passing video tape is being interrupted. Often a loose wire that connects the full-erase head to the VCR's PC board will be the culprit. To remedy this problem, snip off the connector at the full-erase head and hard-solder each end of the two wires to their designated location on the full-erase head and where the wires mount onto the VCR's electrical PC board. This should resolve the fault. Try recording in each of the various record speeds (SP, LP and SLP) to ensure a clean playback picture.

6.24 Various Carriage Assembly Gears

The assembly gears mounted on either side of the typical VCR carriage assembly have been aligned and mounted in place according to the manufacturer's specifications. Many assembly gears have an internal tension spring as a matter of design which applies proper tension on the entire assembly during the insertion and ejection of a tape.

Symptom: *Many times a carriage assembly tension spring will be held in place by a simple nylon tab. Through many hours of normal use, the nylon tabs may break due to stress on the assembly gears. The carriage assembly will not receive a video cassette; or if it does, the video cassette may enter the VCR at a slight angle and get stuck in the process.*

WARNING: As a rule, do not disassemble the carriage assembly out of curiosity, hoping this will correct an original symptom. Often it does not, and the result is a hodgepodge of missing tension springs, loose mounting screws, and misaligned gears.

Toshiba and JVC carriage assembles often incorporate this type of gear. The gear is assembled in three different stages, like a puzzle. It is a good idea to have new parts of the gear on hand and use the faulty gear as an example to ensure that each particular stage is aligned correctly.

6.25 Capstan Motor & Capstan Post

The capstan post is a direct extension of the capstan motor, which is mounted on the underside the VCR chassis. Through many hours of normal use, the capstan post may develop several distinct bands of videotape oxide as a direct result of pulling the videotape through the VCR's tape path. It is vital that the capstan post remain dirt and oil free to ensure that the passing videotape is not damaged and that the playback picture is not distorted.

The capstan motor is responsible for rotating the capstan post at its correct speed while the VCR is in use. A fault may occur if dirt happens to accumulate in the hole through which the capstan post protrudes in the chassis of the VCR. After many hours of use, videotape oxide, dust, and hardened grease may accumulate here and eventually inhibit the capstan motor from rotating at its proper play or record speed.

Symptom: *The symptom for dirt accumulation in the hole at the base of the capstan post is:*
1. *The VCR may stop after being placed in the Play, Review or Record mode.*
2. *If the capstan motor is directly connected to the carriage assembly gears via the drive belt, the VCR may refuse to eject a video cassette since the capstan motor is having difficulty rotating.*
3. *The capstan motor may emit a loud, gut-wrenching drilling sound through the bell of the capstan motor, which is vibrating due to the inhibiting dirt buildup.*

To remedy these problems, it is often necessary to remove the capstan motor from the underside of the chassis. This will allow the accumulated videotape oxide deposits, hardened grease, and dust to be removed from the hole in which the capstan post protrudes though in the VCR chassis. Also, the entire capstan post can be cleaned at this time. Once this area of the VCR has been cleaned, reinsert the capstan motor and test the performance of the VCR in all of its various Play, Review and Record modes. It should continue to work well for many hours of enjoyment.

6.26 Capstan Motor Sensor

Located on the lower side of the typical VCR chassis is a large sensor mounted next to the capstan motor. This sensor is designed to detect a magnetic impulse as the capstan motor rotates. The VCR microprocessor interprets this information to regulate proper rotation speed for the capstan motor. Through many hours of normal use, this sensor may loosen and "drift" away from the side of the capstan motor.

Symptom: The VCR may play a videotape at an accelerated speed that is different from what the videotape was recorded in. It may seem that all videotapes played on the VCR are in Fast-Forward mode.

To remedy this problem, it may be necessary to loosen the mounting screw at the base of the sensor and, while the VCR is playing a videotape, slowly adjust the sensor either forward or backward until the playback audio and picture quality to the video are playing in their correct playback speed.

6.27 Rubber Buffers

VCR manufacturers often use several rubber spacers or buffers between the various parts that come in contact with one another in the typical VCR drive assembly. After many hours of normal use, these rubber buffers tend to lose their rigidity, and the rubber material becomes very soft. When the rubber buffer deteriorates, the space between the two particular parts will vary slightly. Believe it or not, this slight difference in spacing between the two VCR parts can have an adverse affect on the VCR's performance.

The best example of this is in the FUNAI brand VCR format. Located just under the cassette tray, inside the VCR, is a simple rubber buffer. When the wall of this buffer deteriorates through many hours of normal use, the spacing between the drive assembly gears decreases enough for a fault to occur.

Symptom: The gears of the drive assembly do not engage when either the Rewind or Fast-Forward command is selected from the front panel of the VCR.

To remedy this particular symptom, simply replace the rubber buffer with a new one. Also, it is possible to use rubber surgical tubing that is the same width and circumference as the rubber buffer. With the new rubber buffer in place, the VCR will miraculously once again perform well in all of its various modes.

6.28 Traveler Arm Post

During the course of troubleshooting a VCR, it is NOT RECOMMENDED to spray the internal components of the VCR with compressed air. If a small part within the VCR has come unfastened and goes unnoticed upon first inspection, it does not make sense to spray it out the VCR and onto the carpeting where it would most likely be lost for good.

Figure 6-15. A nylon traveler arm. (Courtesy of PRB.)

This could not be stressed more than for the Emerson VCR, which incorporates a nylon transfer arm. (*Figure 6-15.*) This particular transfer arm has a small metal pin that protrudes from the transfer arm into the chassis of the VCR. When the VCR is put into Play mode, this transfer arm is mechanically positioned into place to execute the Play command.

However, since the transfer arm is made of nylon, through many hours of normal use, the nylon arm tends to flex, and the metal pin may loosen and eventually drop out of the hole in the transfer arm. When this occurs, the small transfer pin can be heard rattling across the lower pan when the VCR is tilted from side to side. Keep in mind that this transfer pin will most likely fall out of the VCR onto your work table once the lower pan has been removed.

Symptom: The VCR will not go into Play mode after the Play command has been selected on the VCR's front panel. When the pin falls out, the pinch roller does not make contact with the rotating capstan post, so the videotape does not advance and the VCR defaults to an "all-stop" mode to prevent the videotape from being damaged.

Figure 6-16. Several types of relay switches.

To remedy this problem, simply apply a small amount of bonding cement or glue on the hole in the transfer arm and reinsert the metal pin. After the glue has cured, test the performance of the VCR in all of its various play and record modes. Although this particular symptom tends to occur often in the Emerson brand, any VCR that contains a metal pin in a nylon transfer arm is susceptible to this problem at any given time.

6.29 Relay Switch

Early models of VCRs often incorporate a relay switch as a matter of design. (*Figure 6-16*.) Often the relay switch is housed within a plastic case that has a clear removable cover for accessing the sweep arms of the switch for cleaning. When the VCR is in use, and a command is selected from the VCR's front panel, the sweep arms within the relay switch "toggle" back and forth to electronically execute that particular command. This action of toggling back and forth can result into a fault.

Aside from the normal accumulated dirt which may develop on the contacts of each sweep arm, a sweep arm may vibrate enough through the toggling motion to become stuck in an open position. If this is the case, the relay switch will no longer be able to electronically execute a particular command selected from the VCR's front panel.

Symptom: The symptom for a relay switch sweep arm that is either dirty or stuck in the open position might be:
1. The VCR may accept a video cassette but does not respond to a particular front panel command such as Play, Rewind or Record.
2. If the contacts of the relay switch are dirty, the VCR may exhibit an intermittent "motorboat" sound or a loud whistle while the VCR is in the Play mode.

To remedy this problem, remove the plastic cover to the relay switch and visually inspect the sweep arms to ensure they toggle back and forth with each different command selected from the VCR's front panel. If the VCR enters one mode and not another, chances are a sweep arm is either dirty or stuck in an open position.

To remedy the motorboat sound, it is a good idea to clean the contacts of each sweep arm with a moistened toothpick that has been soaked in cleaning solution, to ensure that the contact on each sweep arm is clean and not bent in any way.

6.30 LED/Lamp

Located between the supply and take-up spool is either a light emitting diode (LED) or an incandescent lamp (lamp), depending on the make and model of the VCR you are working on. (*Figure 6-17.*) For a detailed explanation on the purpose of the LED/lamp, it is recommended that you read section 3.12 in Chapter Three.

Since this area within the VCR can be very frustrating to troubleshoot, understanding the different test techniques for determining a failed LED/lamp with an infra-red light detector will often save valuable diagnostic time.

Symptom: When an LED/lamp fails, the symptom is often:
1. *The VCR will power-up and receive a video cassette, but the only command it responds to is the EJECT command.*
2. *The incandescent bulb often turns burnt gray in color and does not light when the VCR is powered on. The VCR may only respond to the EJECT command.*

To remedy this problem, the failed LED/lamp needs to be replaced with its exact manufacturer's replacement part. Although LED/lamps may look similar in both size and shape, it is not necessarily true that they are interchangeable. Once the new LED/lamp has been installed, test the VCR in all of its various play and record modes. It should continue to work properly for many hours of normal use.

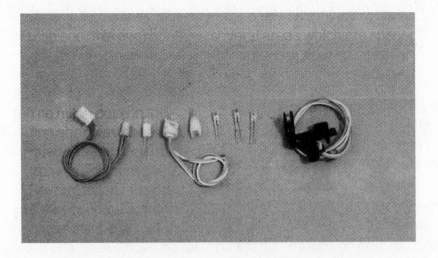

Figure 6-17. *Examples of light emitting diodes (LEDs)*
and incandescent lamps.

Chapter Seven
TWELVE COMMON VCR PART REPLACEMENT PROCEDURES

Chapter Seven
TWELVE COMMON VCR PART REPLACEMENT PROCEDURES

The information that has been provided in the preceding chapters is often all that is required for the novice to successfully repair VCRs. The VCR disassembly steps outlined in the first six chapters enable you to access the various parts within the typical VCR in order to perform a tape path cleaning, replace a drive belt, or change an idler drive tire to rectify an original symptom. However, in the course of troubleshooting a VCR for the root of its symptom, it is often necessary to perform a part replacement procedure to restore the VCR to its proper working order. The information contained in this chapter will advance the novice VCR repair person up to a standard at which most VCR service centers operate.

Starting with a simple power supply fuse and coming full circle to replacing the video heads, after reading this chapter, a novice VCR owner will be able to repair nearly 90% of all problems which affect VCRs on the market today.

If you were unable to rectify the problem with your VCR after reading the first six chapters, chances are you will find the solution in this particular chapter and will be able to apply this information to other symptoms that are referenced within this text.

UNPLUG YOUR VCR NOW!!!

Before continuing with the different replacement procedures, the following list of small hand tools will be referred to throughout the various adjustment or replacement steps to rectify an original symptom. It is very important to use the proper hand tools when adjusting or replacing a particular VCR part, to ensure the VCR is not damaged.

One hand tool in particular that is used often is the soldering iron. The proper technique for using a soldering iron is outlined in Chapter One. It is a good idea to practice soldering and desoldering various parts in a junker VCR before attempting a replacement procedure in a working VCR.

Other small hand tools that are needed include:
    Video head extractor.
    ACE head adjustment wrench.
    Guidepost adjustment wrench.

Solder.
Desoldering braid.
Tweezers.
VCR Allen wrench kit.

These tools can usually be found at your local electronics parts supplier. Also, it is recommended that you purchase manufacturer's replacement parts when performing any of the following repair procedures. By using manufacturer's part numbers, you avoid second-guessing your original repair work and avoid callbacks from angry customers or disgruntled relatives.

7.1 Replacing a Power Supply Fuse

VCR power supply fuses are located at the rear of the VCR in the power supply. Fuses are used within the VCR as a built-in safety feature to protect the various electrical components from a potential power surge through the VCR's power cord. A faulty power supply fuse is often easy to detect since its middle glass portion becomes burnt gray in color when the fuse overheats and fails.

This replacement procedure is very simple and can be done in a few minutes.

Step #1: Locate the Faulty Power Supply Fuse

Power supply fuses are located toward the back where electricity enters the VCR. A power supply fuse contains a thin wire housed within a glass midsection. When the incoming electricity exceeds a specific tolerance, the wire overheats and breaks, and the fuse is rendered useless.

It is often easy to determine which power supply fuse is faulty since the glass midsection tends to become burnt gray in color. Power supply fuses may be hidden from sight beneath plastic covers.

Step #2: Remove the Failed Fuse

Once you have successfully located the faulty power supply fuse, it is a good idea to use a nonconductive plastic or wooden probe tool in order to release the failed fuse from its mounting clips on the VCR PC board. Use a pair of tweezers to grasp the failed fuse and remove it from the VCR. (*Figure 7-1.*)

Step #3: Reinsert the New Power Supply Fuse

Printed on the metal bands located on either end of the power supply fuse are the electrical specifications of the fuse. Often a VCR's power

Figure 7-1. Removing a failed power supply fuse from a VCR.

supply fuse will range between 1.2 - 2.5 amps/125 volts. It is very important that a faulty power supply fuse is replaced with its exact replacement part before the power cord is plugged into a power supply outlet in order to test the performance of the VCR.

Reinsert the new fuse, reassemble the VCR, and test its performance. The power should remain on and the VCR should work in all of its various modes.

7.2 LED/Lamp

Figure 7-2. A light emitting diode (LED). (Courtesy of PRB.)

Located between the take-up and supply spool is either a light emitting diode (LED) (*Figure 7-2*) or an incandescent lamp (*Figure 7-3*), depending on the make and model of the VCR you are currently working on. The LED/lamp should remain in proper working order or the VCR will exhibit a symptom. This particular part is relatively easy to remove from the VCR and requires only several minutes to desolder, remove, and replace.

Figure 7-3. An incandescent lamp. (Courtesy of PRB.)

Step #1: Access the LED/Lamp Lead Wires

The first step in removing a failed LED/lamp is to determine how it is mounted to the chassis of the VCR. Typically, a single mounting screw is used, or a nylon holder is clipped into the metal chassis of the VCR. Once the holder has been disconnected from the chassis, turn the LED/lamp over to access its lead wires.

Step #2: Desolder, Remove and Replace the Faulty LED/Lamp

Using a well-heated soldering iron and a length of desoldering braid, desolder each lead wire. The failed LED/lamp should simply slip out of its nylon holder.

VCRs which contain an incandescent lamp are often connected with a male/female connector. If this is the case, the two lead wires simply unfasten at the connector, and the incandescent lamp can be removed and replaced with a new one.

With the new LED/lamp installed, solder each lead wire in the same order that it was removed if it is of the LED type, or secure the connector if the VCR incorporates an incandescent lamp. Reassemble the VCR and test its performance in all of its various modes. It should continue to work well.

7.3   Program/Timer IC

Many RCA type VCRs tend to use a program IC, which is responsible for storing program/timer information. (*Figure 7-4.*) This stored information enables the VCR to begin recording a televised program at a specified date and time.

Figure 7-4. A typical program IC. (Courtesy of PRB.)

The symptom of this particular IC tends to be very obvious. It resembles either "small houses" or a "pac-man" type maze displayed on the television screen through the VCR:

〔〕

Because this particular IC is responsible for maintaining the program/timer information, it is often located on or around the common PC board directly behind the VCR's front panel. More often than not, the front panel will need to be removed in order to access this IC for replacement.

Step #1: Locate the Failed Program/Timer IC

With the VCR front panel removed, read the code numbers printed on each IC in the location of the VCR command PC board. This is the IC that will be removed from the VCR.

WARNING: *It is a good idea to draw a simple diagram of the failed program/timer IC while it is mounted in the VCR before removing it. Integrated circuits often have a notch at one end in order to distinguish its front from its rear. Your simple diagram of the failed IC will ensure that the new program/timer IC will be installed correctly. The method of drawing a simple diagram and desoldering a failed integrated circuit can be applied to any IC found within the VCR.*

Step #2: Desolder, Remove, and Replace the Faulty IC

Once you have successfully located the faulty program/timer IC, the PC board it is mounted to will need to be lifted from the VCR so that you can desolder the lead wires of the failed IC. This PC board is often secured to the frame of the VCR with two mounting screws, or is simply held in place with nylon locking tabs.

With the proper mounting screws removed, lift up on the PC board that contains the failed program IC in order to access and desolder its lead wires. Using a well-heated soldering iron and a length of desoldering braid, remove the solder from each of the lead wires on the failed program/timer IC.

Once the solder has been removed, the failed program IC can then be replaced with a new one. Install the new IC on the VCR's PC board and solder it into place. Secure the PC board into place and reassemble the VCR. Test the performance of the new program/timer IC. The "pac-man" design should be eliminated, and the VCR should retain programmed information to record at a later time and date.

Alternate Example: Fisher Loading IC #LB 1649

One particular loading IC, part number LB 1649, is located in Fisher brand VCRs. When this particular IC fails, the symptom might be that the carriage assembly refuses to either accept or eject a video cassette. Until this particular IC is replaced with a new one, the VCR will not work properly. Replace this IC using the same procedures outlined above.

7.4 Take-up Spool Sensor

This sensor is located under the take-up spool, described in Chapter Four. (*Figure 7-5.*) It is a good idea to reference this chapter when performing this replacement procedure to ensure another symptom does not occur.

Figure 7-5. A take-up spool sensor. (Courtesy of PRB.)

Through many hours of normal use, this particular sensor may fail. The symptom for a failed take-up spool sensor might be:
1. The VCR immediately stops when the Play command is selected from the VCR's front panel.
2. The digital tape counter does not increment while the VCR is in use, and the VCR powers off shortly afterwards.

It is often possible to detect whether or not a take-up spool sensor is faulty in a VCR that has a digital tape counter. Simply power-on the VCR and rotate the take-up spool by hand. The digital tape counter should increment. If it does not, chances are the take-up sensor is faulty and will need to be replaced with a new one before the VCR will work correctly.

Step #1: Remove the Take-up Spool

The first step in removing a faulty take-up spool sensor is to remove the take-up spool located over it. Remove the nylon locking washer which keeps the take-up spool secure to its mounting post. The take-up spool should simply lift up from its mounting post once the locking washer is removed.

Step #2: Access and Remove Faulty Take-up Sensor

Once the take-up spool has been removed, the next step is to access and remove the faulty sensor. Depending on the make and model of your VCR, it may be necessary to access the faulty take-up sensor from the underside of the VCR chassis.

Using a well-heated soldering iron and a length of desoldering braid, simply remove the solder on each contact of the faulty sensor. Once the solder has been removed, the faulty take-up spool sensor can then be extracted from the VCR.

Step #3: Install the New Take-up Sensor

Reinsert the new sensor that is of correct size and shape for the particular VCR you are currently working on. Solder its lead wires into place and secure the take-up spool onto its mounting post with its locking washer.

Test the performance of the VCR. It should continue to operate in all of its various modes. Also, the tape counter should now increment while the VCR is in use.

7.5 STK Voltage Regulator IC

An important point to keep in mind is the fact that not all VCRs contain an STK voltage regulator IC. However, for the VCR formats that do contain one, STK ICs are located within the power supply. The STK IC is a voltage regulator and is considered a "hot-running" IC. (*Figure 7-6.*)

Figure 7-6. *An STK voltage regulator IC. (Courtesy of PRB.)*

Manufacturers of VCRs apply a liberal coat of heat sink grease to the back of each STK IC. The heat sink grease helps to diffuse much of the heat generated by this IC in an attempt to prolong its useful life. However, after many hours of normal use, this particular IC tends to fail, resulting in a fault. The symptoms for a failed STK IC might be:

1. The upper head assembly does not rotate when the Play command is selected.
2. The power light on the VCR's front panel may blink for several seconds before stopping.
3. The drive or loading motors in the VCR cannot be heard initializing when the power supply cord is plugged into a power supply outlet.

In any case, if the STK IC is faulty, it will need to be replaced with its exact manufacturer's replacement part.

Step #1: Remove Power Supply Mounting Screws

The first step in replacing a failed STK IC is to access its solder contacts on the PC board. In order to do this, the PC board on which the faulty STK IC is mounted will most likely need to be lifted from the VCR. To do this, there are several large chassis mounting screws which will need to be removed first. Often an entire section of the VCR's power supply will need to be lifted as well.

Step #2: Desolder and Remove the STK IC

Once the power supply PC board that contains the failed STK IC has been lifted out of the VCR, turn the PC board over to access and desolder the lead wires of the faulty STK IC. Using a well-heated soldering iron and a length of desoldering braid, desolder each lead wire of the faulty STK IC until it can be easily removed from the power supply PC board.

In many VCR power supplies, the STK IC will be mounted on a metallic heat sink with two mounting screws. This heat sink assists in diffusing the heat generated by this hot-running IC. Remove the heat sink mounting screws so that the faulty STK IC can be removed from the VCR.

Step #3: Apply a Liberal Coat of Heat Sink Grease to the Back of the New STK IC

Before inserting the new STK IC and securing it permanently to the VCR, it is recommended that you coat the back of the new STK IC with heat sink grease. (*Figure 7-7.*) Heat sink grease can be purchased at your local electrical supply store and helps to dissipate much of the heat generated.

Figure 7-7. *Applying heat sink grease to the back of an STK IC.*

Step #4: Reinstall the New STK IC and Solder it Into Place

After the heat sink grease has been applied to the back of the new STK IC, it is a good idea to mount the new STK IC to the metallic heat sink with its mounting screws FIRST. This will keep the new STK IC stationary while the lead wires are soldered. Insert the lead wires of the new STK IC into the power supply PC board and solder them into place.

Step #5: Reinstall the Power Supply PC Board into the VCR

Once the new STK IC has been installed, secure the power supply PC board into the VCR with its chassis mounting screws.

It is a good idea to test the performance of the VCR with the new STK IC installed. The previous symptoms should be eliminated and the VCR should continue to work properly in all of its various modes.

7.6 DC-to-DC Converter

A DC-to-DC converter is often located behind the digital clock display on the front panel of many early RCA brand VCRs. The DC-to-DC converter is a hot-running component and tends to fail after many hours of normal use. When a DC-to-DC converter fails, often the digital clock display will not light when the VCR is powered on. If this is the case, the VCR cannot be programmed to record a televised program at a future time and date, since the time and date function on the front panel display does not work.

Manufacturers of VCRs have mounted a small heat sink on top of newer DC-to-DC converters in an attempt to prolong the useful life of the converter. Also, the new DC-to-DC converter kits include three capacitors, an IC protector, and the DC-to-DC converter itself. When a DC-to-DC converter kit is installed, be certain to install all the parts that are included in the replacement kit. This will ensure that the failed part creating the symptom is removed and replaced with a known working one.

Often, the DC-to-DC converter and its various components will be located within the VCR just behind the front panel. It is a good idea to have the new converter kit on hand when performing this replacement procedure to ensure that the proper components are correctly identified and removed.

Step #1: Lift the PC Board Up and Out of the VCR

The DC-to-DC converter often will be mounted to a PC board located behind the VCR's front panel. This PC board needs to be lifted from the VCR in order to access the solder connections of the faulty DC-to-DC converter and its surrounding components.

Once you have correctly located the correct PC board, disconnect the harness or ribbon cable. This connector is in plain view and can be readily accessed.

Next, lift up on the two sides of the PC board. Use little force when lifting up on this particular PC board since there is a row of connecting pins running along its bottom edge. After lifting this PC board out of the VCR, turn the PC board over to access the solder contacts of the faulty components.

Step #2: Desolder and Reinstall Parts of the Kit

It is a good idea to have the new DC-to-DC converter kit on hand and to install each component one at a time. This is to ensure that the proper components are identified and installed in their correct position on the PC board.

Using a well-heated soldering iron and a length of desoldering braid, remove each component from the PC board and replace it with its new one from the DC-to-DC converter kit.

Step #3: Install the PC Board Containing the New Kit into the VCR

Once all of the various components of the new DC-to-DC converter kit have been installed on the PC board, simply return the PC board to the VCR. Be certain not to bend the row of connectors that run along its lower edge. Also, the harness or ribbon cable will need to be reconnected to complete the repair procedure. Reassemble the VCR and test the performance of the digital clock display. The VCR should now retain program information.

7.7 Carriage Assembly Motor

Nearly all VCR formats on the market today contain a carriage assembly loading motor within the VCR. (*Figure 7-8.*) A symptom can occur if the loading motor fails. Through many hours of normal use, these small motors tend to overheat, their inner contacts freeze up, and the drive shaft stops rotating. When this occurs, the symptom might be:

1. The VCR refuses to accept or eject a video cassette.
2. The loading motor becomes excessively hot while the VCR is in use.

Figure 7-8. Drawing of a carriage assembly loading motor.

If this is the case, the loading motor is most likely faulty and will need to be replaced with a new one before the VCR will receive and eject a video cassette properly. It is a good idea to have several new loading motors on hand since they are so popular, inexpensive, and relatively easy to replace.

The first step in replacing a faulty loading motor is to extract the VCR's carriage assembly. If you have not already removed the VCR's carriage assembly, it is recommended that you read Chapter Three now.

Since the manufacturers of VCRs use a variety of different methods for securing the loading motor to the side of the carriage assembly, it would be impossible to outline the removal procedure for each VCR format in detail. However, a checklist of four specific areas can be applied to every carriage assembly you happen to work on, and should be taken into consideration when attempting to remove a failed loading motor. Keep in mind that the four steps that make up this checklist can be executed in any given order, depending on the design of the carriage assembly in your VCR.

Step #1: Remove Loading Motor Mounting Screws

Loading motors are often secured on the side of the carriage assembly with two small mounting screws. Also, if the loading motor is secured to a mounting plate, it may be necessary to remove the mounting plate first in order to access and remove the mounting screws.

Step #2: Desolder Loading Motor Solder Contacts

The carriage assembly loading motor is connected to the VCR's electrical format through two contacts that provide the loading motor with electricity. Often, there will either be two wires, red (+) and blue (-), or two solder contacts located at the rear of the loading motor. In either case, the solder contacts will need to be desoldered with a well-heated soldering iron and a length of desoldering braid before the faulty loading motor can be removed and replaced with a new one.

Be certain to solder the lead wires to their proper contacts on the new loading motor or a fault may occur when the VCR's performance is tested.

Step #3: Mark the Teeth of Connecting Cam Gears

Many times, when the carriage assembly loading motor is removed, the connecting cam gears may rotate out of alignment. It is a good idea to mark the teeth of the adjacent gear that come in contact with the main cam gear of the carriage assembly, before removing the faulty loading motor. This will ensure that the teeth of the surrounding gears will be correctly aligned with the main cam gear of the carriage assembly once the new loading motor is installed.

Step #4: Release Any Nylon Locking Tabs

Manufacturers of VCRs may use one or more nylon locking tabs to secure the loading motor to the side of the carriage assembly. Often these particular nylon locking tabs need to be released before the faulty loading motor can be removed.

Panasonic brand VCRs often incorporate a nylon locking tab at the rear of the loading motor. This nylon tab is often held tightly against the loading motor with a single mounting screw located directly behind it. Be certain to remove this mounting screw first before attempting to release the nylon locking tab.

Once all of the individual steps in the loading motor removal checklist have been taken into consideration, attempt to remove the faulty loading motor in the VCR you are currently working on. It should come lose without too much difficulty.

Simply reverse the disassembly steps in order to secure the new loading motor into place.

7.8 Guidepost Adjustment

The guideposts play an integral role in the overall performance of the VCR. When the Play or Record command has been selected, the guideposts mechanically extend forward, pulling the videotape out of its plastic cassette shell and applying it to the VCR's tape path.

Through many hours of normal use, the guideposts may loosen and come out of alignment with the rest of the VCR's tape path parts. When this occurs, a symptom may arise.

Typically, the symptom for a guidepost that is out of alignment with the rest of the VCR's tape path parts might be bands of horizontal white lines on either the upper or lower portion of the playback picture. Adjusting the tracking control knob will not correct the problem.

When either the left or right guidepost becomes out of alignment with the other VCR tape path parts, it will need to be adjusted before the VCR will exhibit a clear video playback picture again.

It is recommended that you use a video head speed tester cassette when performing this particular adjustment procedure. A video head speed tester cassette will make it possible to observe the playback clarity and adjust accordingly for each of the different recorded speeds (SP, LP and SLP). See Chapter One, *Tricks of the Trade*, for information on how to construct your own speed tester cassette.

Step #1: Access and Loosen the Base Allen Nut on Each Guidepost

Located at the base of each guidepost is a very small Allen nut that keeps the entire guidepost from spinning freely while the VCR is in use. This particular locking Allen nut MUST be loosened slightly to allow the guidepost to rotate for this adjustment procedure. (*Figure 7-9*.) Using the correct size VCR Allen wrench, loosen the locking nut at the base of each guidepost.

Figure 7-9. *Illustration of a guidepost, showing the location of the Allen nut.*

Step #2: Insert Video Head Speed Test Cassette and Press Play

After the base Allen nut has been loosened slightly on each guidepost, insert the video head speed test tape. Press the Play command and observe the playback picture. You should not be concerned about the quality of the playback picture at this point.

WARNING: *NEVER "tweak" the left or right guidepost without FIRST loosening the Allen nut located at the base of either guide post. If this Allen nut is not loosened beforehand, the guidepost may break off, become permanently stripped, or will continue to rotate while the VCR is in use.*

WARNING: *DO NOT loosen the base locking nut to the point where the guidepost can spin freely. If the guidepost is able to spin freely, the height adjustment will not have any effect since each guidepost will most likely rotate once the test cassette is ejected.*

For this particular adjustment, move the tracking control knob to its center "notch" position, if it has not been centered already. The tracking control knob should remain in its center notch position for the duration of the adjustment procedure.

NOW you should be concerned with the quality of the video playback picture. Most likely there are several bands of white lines running horizontally across the screen at either its upper or lower edge since one or both of the guideposts is out of alignment with the rest of the VCR's tape path parts. If this is the case, both guideposts will need to be adjusted before the VCR will exhibit a clear picture in each of the three play speeds (SP, LP and SLP).

Step #3: Adjust Each Guidepost Accordingly

As the speed test tape is playing in the VCR, begin the adjustment procedure by adjusting the left guidepost first. Using the guidepost adjustment tool, slowly rotate the left guidepost until the UPPER portion of the video playback picture is clear of any white bands or horizontal lines.

Continue with the right guidepost. Using the guidepost adjustment tool, slowly rotate the right guidepost until the LOWER portion of the playback video picture is clear of any white bands or horizontal lines.

Once the upper and lower portion to the playback video picture is clear, it is a good idea to observe the picture quality in each of the play speeds (SP, LP and SLP) as the head speed test tape is being played. If white bands of snow or horizontal lines appear, adjust the guideposts accordingly in order to receive a clear playback picture in each of the play speeds.

Step #4: Eject the Video Head Speed Test Tape and Tighten the Base Allen Nut on Each Guidepost

After adjusting either the left and/or right guidepost, it is necessary to tighten the Allen nut at the base of each guidepost to ensure that the height adjustment for each guidepost remains in the adjusted position. It is best to first eject the video head speed test tape so you can access each Allen nut locking screw. Using the correct size VCR Allen wrench, go ahead and tighten each base locking nut. Be certain not to rotate either guidepost or the adjustment procedure will be futile.

Once the guideposts have been tightened, it is a good idea to test the playback picture quality of each play speed (SP, LP and SLP) by using the video head speed test tape. Enter the video head speed test tape into the VCR, press Play, and observe the picture quality of each particular playback speed. A clear picture should be present for each playback speed if the guidepost adjustment procedure was performed correctly.

7.9 ACE Head Adjustment

The adjustment procedure for adjusting the ACE head assembly is much like the adjustment procedure for the guideposts. The tracking control knob should be at its center "notch" position throughout the course of the adjustment. Also, the playback picture quality should be taken into consideration when the adjustment is being performed.

Step #1: Insert the Video Head Speed Test Tape and Press Play

Once the tracking control knob is at its center notch position, insert the video head speed test tape and press Play. The VCR should load the videotape and begin playing. Observe the playback video picture. Most likely there will be several bands or lines of snow running horizontally across the screen. This is a good indication that the ACE head needs to be adjusted.

Step #2: Adjust the X-Nut at the Base of the ACE Head

Carefully place the ACE adjustment tool on the X-nut located at the base of this head assembly. As the head speed test tape switches between the various play speeds, observe the playback picture quality for each particular speed. Turn the X-nut either to the left or to the right to remove any snow or horizontal lines that appear on the television screen. (*Figure 7-10.*) Continue to do this for each playback speed on the test tape. When every playback speed remains clear and free of snow or horizontal lines, the ACE head is aligned with the rest of the VCR's tape path parts.

Figure 7-10. *Turning the X-nut to adjust the ACE head.*

Step #3: Test the Performance of the Tracking Control Knob

The sign of a good ACE head alignment is denoted by how clear the playback picture remains when the tracking control knob is rotated to the full left or full right. As the video head speed test tape is playing, test your adjustment by turning the tracking control knob to the full left and full right position for each particular playback speed (SP, LP and SLP).

Keep in mind, it is quite normal for the playback picture to dissolve more and more as the tracking control knob is rotated further to the left or right. However, a clear playback picture should always be present once the tracking control knob is set to its center notch position regardless of the recorded play speed.

7.10 RF Modulator

Located on the rear panel of the typical VCR are two posts which protrude from the body of the VCR. These posts are labeled "Antenna In" and "Out To TV". Also, there is a 3/4 channel switch nearby. The protruding posts and the 3/4 channel switch are part of the VCR's radio frequency modulator, or RF modulator.

Like the carriage assembly loading motor, it would be impossible to outline each different method used for mounting the RF modulator in VCRs. However, it is possible to create an RF modulator removal checklist that can be applied to every VCR you happen to work on. The contents of this RF modulator checklist can be executed in any particular order depending on the accessibility of the individual steps based on the make and model of your VCR.

For example, Fisher brand VCRs may require removal of a mounting screw before a lead wire can be desoldered, while a Zenith brand VCR may require the desoldering of a lead wire before a mounting screw can be removed.

Step #1: Desolder the Lead Wires

Often there will be several wires soldered to the RF modulator that need to be desoldered in order to extract the faulty RF modulator. It is a good idea to draw a simple diagram of these particular wires before desoldering them from the faulty RF modulator. This diagram can then be used to secure the wires in their proper location on the new RF modulator once it is installed. The connecting wires are often denoted with a B+, GRN, and B-. Be certain to solder the correct wire to its proper contact when installing the new RF modulator, or a fault may occur.

Step #2: Disconnect One or Two RCA Cables

Aside from the individual wires, there is often an RCA coaxial cable exiting the RF modulator that is connected to the VCR. This particular cable is connected through an RCA jack. To remove the faulty RF modulator, this particular cable must be disconnected as well. Often, a spot of solder is used to keep the RCA connector securely in place. This solder will need to be removed before the cable can be unfastened.

Step #3: Remove Any Internal/External Mounting Screws

Manufacturers of VCRs may use one or more mounting screws to secure the RF modulator into place within the VCR. Be certain to inspect the rear panel as well as the top side of the VCR for these particular mounting screws, which are often red in color to signify their importance. They need to be removed before the faulty RF modulator can be extracted from the VCR.

Step #4: Remove Any External Stereo Post Locking Nuts

When attempting to remove a faulty RF modulator, you will need to inspect the external posts of the RF modulator itself. Often, two stereo locking nuts will be used to keep the mounted RF modulator securely in place within the VCR's format. Never attempt to extract a faulty RF modulator until this area has been examined.

Stereo locking nuts can be used on an RF modulator that has a loose "Antenna In" or "Out To TV" post. The stereo locking nuts will provide added rigidity to the post and may correct a symptom as well as save the expense of having to replace an entire RF modulator.

Step #5: Release the RF Modulator Locking Tabs

The final and often the most frustrating aspect or removing a faulty RF modulator is the fact that the manufacturer has incorporated small locking tabs into the design of the RF modulator. These very small locking tabs are often hidden from sight and are very difficult to access. The locking tabs should be released or opened before the faulty RF modulator is extracted from the VCR.

Because it is impossible to describe the differences between modulator locking tab locations in the various brands of VCRs, it will be up to you to find a creative way to release each of them without incurring damage to the surrounding electrical PC board.

Step #6: Replace Faulty RF Modulator

Once the faulty RF modulator has been extracted from the body of the VCR, simply insert the new RF modulator and reverse the disassembly steps outlined above.

There isn't really a set pattern to removing and installing a faulty RF modulator. However, if the above steps are taken into consideration for each faulty RF modulator you happen to replace, you should fare well in this particular procedure.

7.11 Mode/State Switch

It is very rare that a VCR does not contain a mode/state switch. (*Figure 7-11*.) When a particular command is selected from the VCR's front panel, the mode/state switch is put to work. Through many hours of normal use, the mode/state switch tends to accumulate dirt or fails. If this is the case, the mode/state switch will need to be removed in order to either clean its inner contacts or replace it with a new mode/state switch.

Figure 7-11. *Illustration of a typical mode/state switch.*

Like the carriage assembly loading motor and the RF modulator replacement procedures, there is also a checklist that can be applied to any mode/state switch you happen to remove. The steps outlining this removal procedure can be executed in any given order, depending on the accessibility of the mode/state switch of your VCR.

Step #1: Draw a Color-Coded Wire Diagram

It is always a good idea to draw a simple color-coded diagram of the wires that connect the mode/state switch to the VCR, before desoldering them from the mode/state switch. This simple wire diagram can save valuable time and eliminate a lot of guesswork when securing the wires to the new mode/state switch after it has been installed. The more detailed your diagram is, the easier it will be to reconnect the wires to the new mode/state switch. Be certain to include an overall diagram of the mode/state switch as it appears mounted within the VCR, as well as the colors of the individual connecting wires.

Step #2: Mark Adjacent Gears and Remove Any Obstructions

Often a VCR's mode/state switch will be imbedded within several layers of gears, or located under a loading motor assembly. If this is the case, several obstructions may need to be cleared before the mode/state switch can be accessed and extracted.

It is a good idea to mark the teeth of any adjacent gear that come in contact with the mode/state switch with a pencil. This will ensure that the surrounding gears of the assembly are correctly realigned with the gears of the new mode/state switch as it is being installed.

Step #3: Desolder Wire Contacts

Once you have drawn your color-coded wire diagram and have marked the adjacent gears, desolder the individual wires from the mode/state switch. Simply use a well-heated soldering iron with a length of desoldering braid and proceed slowly. Be certain not to touch the teeth of the adjacent gears with the hot tip of the soldering iron.

Step #4: Remove Mounting Screws

After the lead wires have been desoldered, remove any mounting screws which may be securing the mode/state switch to the chassis of the VCR and insert the new mode/state switch.

It is a good idea to secure the new mode/state switch to the chassis of the VCR with the mounting screw before attempting to solder the lead wires to the new mode/state switch. The mounting screw will keep the new mode/state switch in place and free both of your hands so that you can solder the connecting lead wires.

Once the new mode/state switch has been installed, reassemble the VCR and test its performance. It should continue to work in all of its various modes without hesitating or acting "confused."

7.12 Replacing the Video Heads

All VCRs on the market today have replaceable video heads. Think of the video heads as the "eyes" to your VCR. Without them, the VCR would be unable to read or record videotape information.

After many hours of normal use, or only a few minutes of improper care, the video heads may wear down, chip off, or become so imbedded with dirt that they no longer function properly. If this is the case, the video

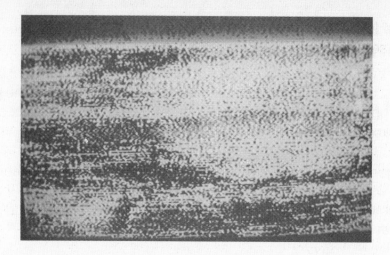

***Figure 7-12**. The TV screen should show a clear playback picture of all of the VCR's play speeds, rather than the distortion shown here.*

heads will need to be replaced in order for the VCR to exhibit a clear playback picture in each of its various play speeds (SP, LP and SLP). (*Figure 7-12.*)

Video heads are a very sensitive area in the VCR and should only be cleaned by hand. For this reason, use of a retail video head cleaner is not recommended. Retail video head cleaners tend to accelerate the wear and tear of the video heads, though they may seem to be a quick fix.

It is the upper head assembly containing the video heads that is actually being replaced. The upper head assembly is located toward the center of the VCR chassis, and is mounted at a slight angle to a lower drum

***Figure 7-13**. The mounting screw that holds the static guard in place.*

motor which is housed within the lower half of the assembly. The upper head assembly can rotate freely and is secured to the lower drum motor with two mounting screws and several solder connections.

Step #1: Remove Any Obstructions

The first step in removing a VCR's upper head assembly is to remove any obstructions which may inhibit the removal process or impede your work. Often a static guard will be mounted to one side of the upper head assembly as a matter of design. This static guard is secured into place with a single mounting screw, and will need to be removed. (*Figure 7-13.*) Other obstructions which may impede your work might include metal head shields, PC boards, and plastic dust covers.

Step #2: Remove the Upper Head Assembly Mounting Screws

All VCR upper head assemblies are secured to the lower drum motor with two mounting screws. (*Figure 7-14.*) Use a free hand to keep the upper head assembly stationary when removing these mounting screws. Do not use force to loosen the mounting screws or the lower drum motor post may come out of alignment. Once the two mounting screws have been removed, set them aside in a safe place away from your work area.

Step #3: Draw a Wire Diagram or Denote Half Circles

If you have not done so already, draw a color-coded wire diagram of the center upper head assembly. The more detailed and accurate your

Figure 7-14. Mounting screws that secure the upper head assembly to the lower drum motor.

diagram, the better chance you have of correctly soldering the lead wires to their proper location on the new upper head assembly.

If the lead wires are not returned to their proper location on the new upper head assembly, a symptom will arise. The symptom might be that one half of the playback picture will be clear while the other half is distorted.

On Panasonic brand VCRs, the upper head assembly is often divided into a white and green half. Mark the lower drum motor post of the green half with a half circle before the faulty upper head assembly is removed. When the new upper head assembly is installed, be certain that the green side aligns with the half circle on the lower drum motor post. This will ensure the new upper head assembly is mounted correctly onto its lead wires.

Step #4: Desolder the Lead Wires from the Upper Head Assembly

Be certain to use a well-heated soldering iron and a length of desoldering braid to remove all of the solder from each lead wire. (*Figure 7-15.*) You have nothing to loose and everything to gain by taking your time at this stage in the removal procedure.

Upper head assembly lead wires can often be disconnected with a pair of tweezers once the solder has been sufficiently heated. However, it is vital that all of the solder be removed from the lead posts if it is of the type which protrudes through the surface of the upper head assembly. Often, these particular solder contacts will be denoted by a "VV."

Figure 7-15. *Use desoldering braid to remove solder from the lead wires.*

Figure 7-16. *The upper head assembly can be removed once separated from the lower drum.*

Step #5: Use a Video Head Extractor to Remove the Faulty Upper Head Assembly

Manufacturers of VCRs have taken into consideration the fact that video heads eventually wear out and need to be replaced. For this particular reason, the upper head assembly has two threaded eyelets in order to receive a video head extracting tool. A video head extracting tool should always be used when either removing a faulty upper head assembly or installing a new one. By utilizing this tool, the risk of damaging the new upper head assembly or the axle of the lower drum motor is kept to a minimum. Also, by using an extracting tool, the surface of the new video heads will not be contaminated by the oil from your fingertips.

Simply fasten the video head extracting tool to the faulty upper head assembly by securing each fastener into its threaded eyelet. Once the fasteners of the video head extracting tool have been secured into the threaded eyelets, slowly rotate the handle of the extracting tool.

Once the faulty upper head assembly begins to separate from the lower drum motor assembly, continue to slowly rotate the handle of the extracting tool until the faulty upper head assembly is free and clear of the VCR. (*Figure 7-16.*)

Step #6: Insert the New Upper Head Assembly and Secure it into Place with its Mounting Screws

Once the faulty upper head assembly has been removed from the VCR, disconnect the head extracting tool and secure it onto the new upper head assembly. This can often be accomplished while the new upper head assembly is seated in its original packaging, in order to avoid touching the upper head assembly with your fingertips.

With the head extracting tool secured in place, hold the new upper head assembly over the lower drum motor and gently lower it down onto the proper lead posts. Continue turning the handle of the head extracting tool until the new upper head assembly is seated flush to the lower drum motor.

It is very important to secure the new upper head assembly into place with its mounting screws before moving onto the next step. If you have not already done so, do this now.

Step #7: Secure the Various Wires to their Proper Contacts as Outlined in the Wire Diagram

Once the new upper head assembly has been seated and secured into place with its mounting screws, solder the individual wires or lead posts into place. Using a well-heated soldering iron, a length of solder, and your color-coded wire diagram, continue around the entire inner circle of the new upper head assembly and secure each solder contact into place.

It is very important that each wire is secured on the new upper head assembly in its correct location. If two wires happen to be crossed, a fault would most likely occur. The symptom for a new upper head assembly which has one or more wires soldered improperly might be that either the upper or lower portion of a playing video tape will be clear while the other half will be snow.

Step #8: Reassemble the VCR and Test the Performance of the New Upper Head Assembly

Once the new upper head assembly has been installed and its lead wires correctly soldered into place, reassemble the VCR and test the performance of the new video heads.

It is recommended that the VCR record several minutes of broadcast programming in each of the record speeds (SP, LP and SLP). Once this has been done, rewind the videotape and play the recording you have just made. The VCR should switch between the various play speeds automatically, exhibiting a clear picture in each.

This completes the repair procedures for this chapter. It is very important to keep in mind that these procedures can be applied to nearly every VCR on the market today.

Chapter Eight
COMMON HUMAN ERRORS IN VCR REPAIR

Chapter Eight
COMMON HUMAN ERRORS IN VCR REPAIR

Human errors are composed of the most common mistakes that novice VCR repair and cleaning people make while attempting to resolve symptoms in their VCR. If any of these human errors goes unnoticed, the result is usually a symptom which did not exist before, leading you to believe that your original repair attempt was not successful.

There is nothing worse than creating another symptom in an attempt to remedy the original problem affecting your VCR, which is why each of these human errors should be considered throughout the entire repair and cleaning procedure. One or more of these particular errors could arise at any point during the troubleshooting, cleaning, or repair process if you are not careful.

If another symptom does occur, DON'T PANIC. Always consider the possible human errors before assuming that your VCR has incurred another, more serious problem. Often you will be relieved to find that a simple drive belt, tension spring, or harness cable has been left unfastened.

8.1 Human Error #1: Overhead Work Light is Too Bright

When you are testing the performance of a VCR before or after it has been serviced, and the lid has been removed, be certain the overhead work light is neither too bright nor too direct.

As a matter of design, the typical VCR carriage assembly has a light-sensitive sensor mounted on both its left and right side. (*Figure 8-1.*) This sensor responds to the infrared light that travels through the body of the video cassette shell once the videotape has reached its end or beginning. The VCR microprocessor reacts to this light by issuing an all-stop command to prevent damage to the videotape. An overhead work light such as a pen light or reading lamp may have the same effect as the infrared light. When this occurs, and you are trying to test the VCR by playing a tape, the VCR will either stop playing if it is in the Play mode, or it will not go into the Play mode at all and will rewind the tape instead.

Figure 8-1. A carriage assembly light sensor. (Courtesy of PRB.)

Figure 8-2. *An unhooked harness or ribbon cable needs to be reconnected before the VCR will work properly.*

To remedy this light-sensitive symptom, simply use a large piece of cardboard to temporarily cover the top of the open VCR. This technique will allow you to view the mechanical workings of the VCR as it is being tested without it randomly stopping or rewinding.

8.2 Human Error #2: Unhooked Harness or Ribbon Cable

Many times you will need to remove the carriage assembly from the VCR to fully access the tape path to replace, adjust, or clean it. Often, when returning the carriage assembly to the VCR to test its performance after your cleaning procedure, the connecting ribbon or harness cable may be left unhooked as an oversight . (*Figure 8-2.*)

When this occurs, the symptom might be that the VCR may power-on all right, but the carriage assembly will not accept a video cassette. Also, the capstan post and take-up spool may rotate simultaneously for several seconds before the VCR powers off.

It is always a good idea to leave an unhooked harness or ribbon cable in plain view when removing the carriage assembly from the VCR. This will ensure that you do not forget to reconnect it once you are finished with your repair or cleaning procedure. You may save yourself hours of misguided "tweaking" in an attempt to remedy a simple symptom.

8.3 Human Error #3: New VCR Belts are Either Too Tight or Too Loose

As a rule, it is very important to replace a faulty VCR drive or loading belt with a new one which is 5% smaller in circumference than the original belt. This accounts for any stretching that occurs over time through normal use, and ensures that the new belt fits correctly.   Make a point to use manufacturer replacement parts that are of correct size in order to avoid second-guessing.

Since VCR belts come in many different shapes and sizes, novices to VCR repair and cleaning are sometimes tempted to use any belt (including rubber bands) that "seems" to fit. However, a symptom may occur if the new VCR drive belt is too loose or too tight. (*Figure 8-3.*)

If a drive or loading motor labors excessively due to the tightness of a new belt, a power supply fuse may fail when a video cassette is entered into the VCR or when a particular command is selected from the VCR's front panel.

Symptom*: The symptom of a VCR belt which is too loose might be:*
1. *The VCR exhibits an "underwater" sound quality in the audio portion of a playing videotape.*
2. *The VCR may attempt to execute a particular command but suddenly powers-off.*

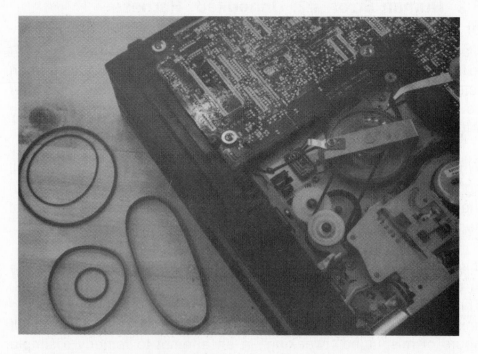

Figure 8-3*. Be sure a new drive or loading belt is the right size before installing it in the VCR.*

Figure 8-4. *A timer switch left in the "On" position.*

3. The VCR may eat videotapes while in the Play or Fast-Forward
    mode.
4. A loud "squelching" noise may be heard when the VCR is
    powered on.
5. A still frame will be present on the playback picture when the
    VCR is playing a videotape.
6. Intermittent patches of snow may randomly appear then disap-
    pear on the playback picture when the VCR is playing a video-
    tape.

8.4 Human Error #4:
Timer Switch is in the "On" Position

Nearly every VCR on the market today requires the timer switch to be set
in the "On" position in order for the VCR to begin recording a preprogrammed television show. It is very easy for the timer switch (located
on the front panel of the VCR) to be accidentally left in the "Timer On"
position during VCR servicing. (*Figure 8-4*.) This human error often
occurs during the disassembly or reassembly stage when the front panel
is being removed or reseated on the chassis.

The symptom of a timer switch that has been accidentally moved or left in
the "On" position is that the VCR may power up but the carriage assembly
will not accept a video cassette. Also, the VCR may not respond to any
front panel commands if it actually does receive a video cassette.

To recognize this human error, the "Timer" light will either be on or it may
be blinking. A VCR tested with its timer switch in the "On" position may

lead you to hours of unnecessary disassembly and reassembly, and misguided "tweaking."

8.5 Human Error #5: A Tension Spring is Left Unhooked

Nearly every VCR on the market today uses one or more tension springs to keep mechanical parts functioning properly. Often, for example, there will be a tension spring connecting an idler assembly wheel to its mounting post. This particular tension spring keeps the idler assembly wheel in constant contact with a drive motor that is often mounted to the underside of the chassis with two mounting screws. This tension spring needs to be unhooked in order to extract the assembly wheel from the VCR to replace the idler drive tire. If this particular tension spring is left unhooked and the VCR is reassembled and its performance tested, a symptom will occur.

Symptom: The symptom of a tension spring that has been left unhooked might be:
1. *The VCR may sound as if it is executing a particular front panel command but then powers off.*
2. *The audio or video portion of a playing videotape may exhibit an "underwater" sound quality.*
3. *The VCR does not advance the videotape when Play, Rewind or Fast-Forward is selected.*

8.6 Human Error #6: The VCR's Front Panel is Not Seated Correctly

More often than not, when servicing VCRs, you will need to remove the front panel in order to disassemble the VCR to access its dirty or faulty parts. When reassembling the VCR, it is very important to seat the front panel correctly in place. Be certain the tracking control knobs, stereo slide posts, timer switches, and tape speed (SP, LP and SLP) switch are not inhibited by the molded plastic of the front panel in any way, or a fault may occur.

Symptom: The symptom of a VCR front panel that is not seated correctly might be:
1. *The timer switch or tracking control knob cannot be moved into the proper position.*
2. *A certain tape speed (SP, LP or SLP) cannot be selected when setting the recording speed.*
3. *The VCR does not respond to a particular command selected from either the front panel or its remote control.*
4. *The tracking control knob does not rotate smoothly when adjusting the playback picture.*
5. *The audio portion of a VCR with stereo recording abilities cannot be regulated from the front panel.*

8.7 Human Error #7: The VCR is Tested Unassembled

As a rule, when servicing VCRs, a VCR's performance needs to be tested once a failed part has been cleaned, replaced or adjusted. The time required to troubleshoot your VCR is often tied up in the disassembly and reassembly process. This is also the time in which a new symptom can arise if caution is not used.

The VCR lid, lower pan, and front panel do not necessarily have to be reinstalled before testing the performance of a repaired VCR. However, it is vital that all of the mechanical parts that were removed during a standard cleaning/repair procedure be correctly returned before testing your VCR. If a mechanical part is not returned to the VCR's format, and the VCR is tested, damage may occur to either the VCR itself or to the videotape being played within the ailing VCR.

For example, VCR formats which contain the track & groove idler drive assembly often incorporate a cassette brake release bracket as a matter of design. When a video cassette is entered into the VCR, the post on the bracket enters a hole located on the underside of the incoming video cassette shell and releases an internal break lever. When this break lever is released, both the supply and take-up spool can rotate freely while the VCR is in use. However, if the break release bracket is not returned to the VCR (as an oversight during the reassembly process), and the performance of the VCR is tested, a fault will occur.

Symptom: The symptom of an integral part that is omitted from the tested VCR might be:
1. The VCR may attempt to execute a particular front panel command but then immediately powers off.
2. The video cassette spools are unable to rotate freely and the videotape snaps in two.
3. The VCR exhibits a loud grinding sound and the teeth on a nylon gear become stripped.

8.8 Human Error #8: Wrong RF Modulator 3/4 Channel Switch Setting

On the back of every VCR on the market today are two protruding posts. These posts are part of the VCR's radio frequency modulator, or RF modulator, which is secured within the frame of the VCR. The RF modulator is responsible for translating information that enters the VCR into "readable" video information, and making this information available to either the channel 3 or 4 setting via the connected television set. Located on the RF modulator is a 3/4 channel switch for this particular purpose. It is very important that this switch is set to its proper channel setting before assuming that your VCR has acquired a new, more serious symptom. Nearly all VCR user manuals suggest designating channel 3 as the outgoing channel from the RF modulator to your television set.

Keep in mind that your television must also be set to this particular RF channel in order to view videotapes and broadcast signals received through your VCR. A simple phrase to keep in mind when the RF modulator and television are set on the same channel is, "The picture quality you see through your VCR (via the television) is the same picture quality that will be recorded." If the RF modulator 3/4 channel switch is accidentally moved from channel 3 to channel 4, and the designated television viewing channel is set on 3, a fault will occur.

Symptom: The symptom for this mix-up in outgoing signals from the VCR's RF modulator might be:
1. *Horizontal lines of snow or fuzz on the playback screen when a videotape is played.*
2. *There is audio but no video when a videotape is played.*
3. *Broadcast signals recorded through the VCR are "scrambled" when played.*

8.9 Human Error #9: Missing Nylon Washer or Mounting Screw

Nearly every VCR on the market today incorporates several cut nylon locking washers and mounting screws to secure various tape path parts on the VCR's chassis.

There are two types of nylon locking washers: "cut" and "whole." The cut nylon washer is reusable since it can be removed without the shape of the washer being affected. However, the whole nylon washer should not be reused after being removed since it must be pried off from the mounting post. This alters the original shape of the washer and may result in a fault once the VCR has been reassembled and is in use.

Also, if an important mounting screw is left out of the VCR, and the VCR's performance is tested, this may result in a symptom which might be:
1. The VCR may seem to be in a particular mode but the videotape is not advancing and the playback picture is not present on the television screen.
2. The VCR exhibits a loud rattle while in any mode.

8.10 Human Error #10: Wrong Size Idler Drive Tire

Located between the supply and take-up spool is a rubber idler drive tire, which plays an integral role in the VCR's overall performance. This

Figure 8-5. *New idler tires must be the right size for the VCR or a serious fault may occur.*

particular VCR part tends to contribute to nearly 75 % of all VCR faults when it either runs smooth or no longer functions properly.

It is vital that the idler drive tire is replaced with its exact manufacturer's replacement part, which is of correct size and shape. (*Figure 8-5.*) Novices to VCR repair and cleaning often use any size idler drive tire at hand so long as it seems to fit well.   If the width and circumference of the new idler drive tire is not correct, a fault may occur which did not exist before. This new symptom might be:

1. The VCR eats videotapes while in use.
2. If the new idler assembly tire is too large, the assembly wheel will bind or "jam up" between the take-up and supply spools, prohibiting the videotape from advancing.
3. If the new idler assembly tire is too small, proper traction will not be made between either spool, and the VCR will often power off several seconds after being placed in a particular mode.

Chapter Nine
COMMON VCR MODELS & THEIR HOT SPOTS

Chapter Nine
COMMON VCR MODELS & THEIR HOT SPOTS

Beginning in the mid-to-late 1970s, the purchasing public had a choice between buying either a Beta or VHS format VCR. Since that time, the Beta format has all but become extinct while the VHS format has become the norm nationwide. The extinction of the Beta format and the standardization of the VHS format has made it possible for the average VCR owner to acquire enough knowledge to maintain the many different brands of VCRs on the market today.

Knowing the model of VCR and its mechanical format is a very important aspect to take into consideration when troubleshooting it. For example, the mechanical format of Fisher VCR is very different from that of Panasonic brand VCRs, though they are both VHS standard. Because mechanical variations exist among the many different brands of VCRs, it is very common to see the same faults again and again in many VCRs of the same model.

This chapter makes an attempt to outline some very common mechanical formats found in VCRs on the market today. As you will see, each model tends to have its own method of accessing the various parts, as well as its share of potential hot spots which often result in a fault after many hours of

Figure 9-1. The interior of a Panasonic model VCR.

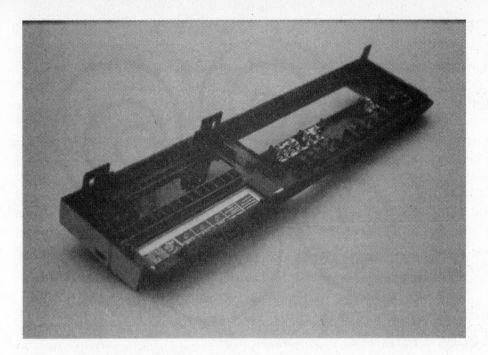

Figure 9-2. *Removed front panel of a Panasonic VCR.*

normal use. Every procedure outlined in this chapter should sound familiar, since we have gone over them in the preceding chapters. If you have not yet read chapters One through Eight, please do so now.

Keep in mind, as the different VCR models are outlined, the level of complexity in disassembling the VCR to correct a symptom becomes increasingly more difficult. Also, if a particular repair procedure seems

Figure 9-3. *Removed Panasonic carriage assembly.*

beyond your ability to correct, don't attempt it. It may be best left to a reliable VCR service center. If possible, ask the service center to describe the repair procedures so that the next time you see the symptom you will understand how to correct it and be able to avoid an inflated repair bill. It will also help to obtain manufacturer's repair documentation or a service manual when examining a particular VCR. Your local service center will be able to show you how to obtain such manuals. Also, not all VCR model numbers are given here. It is assumed that you will be able to recognize the type of VCR you are working on from the description given, since you should already be familiar with identifying the inner workings after reading the preceding chapters.

So, let's consider some of the most popular VCR models in use today and their potential hot spots.

9.1   Panasonic, Sylvania, GE, Quasar; Series Prefix PV, VC, 1VCR, VH

The first of the various VCR models outlined in this chapter is the Panasonic brand, which has a format commonly used in other brands. This particular format was first introduced in the mid-1980s and is one of the most reliable on the market. It is very straightforward to work on. Rumor has it that a VCR of this format can last twenty years if you understand how to replace the various parts that tend to fail. (*Figure 9-1.*)

Figure 9-4. *Inside of the carriage assembly, showing the loading motor. An exact replacement part can be ordered by using the part number of the original part.*

The front panel of a Panasonic brand VCR incorporates a hinge system. To remove this type of front panel, simply release the three upper locking tabs, and the front panel will literally swing forward from the VCR into your hand. (*Figure 9-2*.)

The carriage assembly can easily be extracted from the VCR once its mounting screws have been removed. (*Figure 9-3*.) The mounting screws are often red in color to signify their importance, and tend to be located at the base of the carriage assembly as well as along the upper left and right edges.

It is a good idea to have several compatible loading motors on hand since this brand is very popular, and the loading motor tends to fail after many hours of normal use. (*Figure 9-4*.)

The idler drive assembly is of the pop-out type, and can be easily removed by simply releasing a tension spring and removing the nylon locking washer used to secure the assembly wheel to its mounting post. Often, a standard cleaning may include but not be limited to replacing the VCR's idler drive tire, changing the various loading and drive belts, and cleaning the tape path and video heads.

The upper head assembly (*Figure 9-5*) is often divided into a white half and a green half. It is very helpful to denote the green half when replacing the video heads by drawing a half circle on the lower drum motor post with a pencil. This will ensure that the green half on the new upper head assembly will be aligned and positioned correctly onto its lead posts.

Figure 9-5. *An upper head assembly.*

Also, there is a lower side drive and a loading belt, which are easily accessible for either cleaning or replacing. (*Figure 9-6*.) A simple mounting bracket must be removed before the large drive belt can be accessed, while the smaller loading belt can be replaced by using a pair of tweezers.

Many times, the mode/state switch is easily accessible for replacing since it is mounted on the lower side of the VCR's chassis in plain view. It is always a good idea to draw a color-coded wire diagram of the connecting wires before desoldering them from the faulty mode/state switch. The more detailed your wire diagram, the more likely the new mode/state switch will be reconnected correctly.

Figure 9-6. *The lower side drive and loading belts.*

9.2 Sharp; Series Prefix VC

This particular VCR model, manufactured for Sharp, is very reliable since many of the mechanical parts are constructed of metal. It is very rare that a VCR of this format will need a lot of attention when a symptom occurs. A standard belt replacement, an idler tire change, or a video head cleaning is often all that is required to return this VCR to its proper working condition.

The idler drive assembly is of the mounted type, and is secured into place with two mounting screws that keep the drive assembly secure to the chassis. It is very important to release the lower side drive belt first before extracting the two mounting screws. Once the idler drive assembly has been removed, the rubber idler drive tire simply peels off the perimeter of its plastic assembly wheel for replacement. (*Figure 9-7*.)

Figure 9-7. *A replacement mounted idler drive assembly. (Courtesy of PRB.)*

The carriage assembly removes easily from the VCR once the two base mounting screws have been removed. Often, a ribbon cable will need to be released from the carriage assembly in order to free it from the VCR.

It is a good idea to mark the leading edge of the ribbon cable with a pencil before releasing it from its connector on the side of the carriage assembly. This will ensure that the ribbon cable is inserted correctly back into its

Figure 9-8. *The take-up spool, showing the pinwheel design that helps trigger the sensor functions.*

connector once the carriage assembly has been returned to the chassis. After the VCR's carriage assembly has been removed, many of the various parts of the VCR's tape path are readily accessible for a good cleaning or replacement.

The LED can be removed from the chassis by turning its nylon mounting clip and releasing the female connector from the male connector mounted on the PC board.

The take-up spool sensor lead wires are accessible only from the lower side of the chassis for desoldering. However, the take-up spool sensor itself must be removed and reinserted from the top side of the VCR. (*Figure 9-8.*)

Located on the left-hand side of the chassis towards the rear of the VCR is a small drive belt which tends to go unnoticed. This drive belt connects the loading motor to a set of lower side cam gears. When it needs replacing, it is a good idea to use a pair of tweezers with tips that are bent over one another, in order to slip the drive belt over the lower pulley. Once the drive belt is in place on both of the pulleys, simply open the pair of tweezers and remove it from around the belt.

The VCR's mode/state switch is often located on the lower side of the chassis, imbedded within the cam gears. It will be necessary to first remove several mounting screws to access the mode/state switch in order to clean its contacts or replace it if it is faulty.

Figure 9-9. Removal of the PC board headshield.

Keep in mind that it is a good idea to mark the teeth of the surrounding gears, as well as any potential sliding brackets, with a pencil before lifting the gear assembly out of the VCR. The more detailed your original markings, the better chance you have of correctly realigning the various gears, travel arms and sliding brackets when reinstalling the gear assembly.

9.3   NEC, Toshiba

NEC and Toshiba VCR models are often considered high-end, or (in other words) high-performance. This particular model tends to have several obstacles to overcome before you can access any faulty parts. It is very important to proceed slowly when disassembling this particular format to ensure that another symptom does not occur once the VCR has been reassembled.

After removing the three top-side lid mounting screws and the VCR lid itself, there is often a PC board headshield secured to the carriage assembly with two or more mounting screws. (*Figure 9-9.*) Before lifting up on this particular headshield, be certain to release any potential nylon locking tabs.

Once the PC board headshield has been lifted and secured in its upright position, there is often a metal headshield that needs to be removed as well. (*Figure 9-10.*) This metal headshield is usually secured to the carriage assembly with mounting screws. Simply remove these mounting screws so the metal headshield can be lifted from the VCR.

Figure 9-10. *Removing the metal headshield.*

Located towards the rear of the chassis, in plain view, is a small drive belt that can now be accessed, removed, and replaced by hand. It is a good idea to carry this particular size belt as a stock item since nearly all NEC VCRs incorporate this drive belt.

To access the loading belt on the right-hand side of the carriage assembly, as well as access the idler drive tire, you will need to remove the carriage assembly. In order to do this, however, the front panel must first be removed. Often, the front panel will be secured to the frame of the VCR with several large mounting screws, located around the perimeter of the front panel. Remove these mounting screws and the front panel will be released from the frame of the VCR.

After the front panel has been removed, the carriage assembly is now accessible and can be removed. Located on the inner rear corners of the carriage assembly are two mounting screws which secure the carriage assembly to the metal chassis. These mounting screws are accessible by inserting a screwdriver through the access holes located on either side of the carriage assembly. It is a good idea to use a pair of tweezers to hold each mounting screw in place as it is being removed, so it won't fall into the VCR.

Once the carriage assembly has been removed and the ribbon cable disconnected, there may be a small loading belt located on the right-hand side of the carriage assembly that can now be accessed for replacement.

As a rule, it is a good idea to remove the old loading belt from around each pulley and secure the new loading belt last. This will reduce the amount of stress being applied to the mechanism. To remove this loading belt, a small nylon locking clip that keeps the drive axle secured to the side of the carriage assembly must first be released. Don't worry if the carriage assembly gears rotate slightly during this procedure. Once the nylon clip has been refastened and the new loading belt installed around each pulley, the drive axle can then be rotated to initialize the position of the various carriage assembly gears to the "full eject" position.

NEC VCRs often use a simple pop-out idler drive assembly. (*Figure 9-11.*) Remove the nylon locking washer, and the idler assembly can then be lifted up from its mounting post so that you can replace the idler drive tire.

Figure 9-11. A replacement pop-out idler drive assembly. (Courtesy of PRB.)

Through many hours of normal use, the guideposts tend to loosen and begin to rotate. It is not that the base locking nut has loosened (which is usually the case in most other VCRs), but that the entire post tends to drift up and down within its metal base. The symptom for a drifting guidepost is a series of horizontal lines that appear on either the top or bottom of the playback picture. (*Figure 9-12.*) Adjusting the tracking control knob will not correct the problem.

To remedy this problem, it will be necessary to push the guidepost up from beneath the chassis and adjust its height accordingly while playing a

Figure 9-12. Drifting guideposts often result in horizontal lines appearing across the playback picture.

Figure 9-13. *Replacement NEC carriage assembly gear.*

video head speed test tape. Adjust the height of the drifting guidepost until a clear picture is present in each of the three playback speeds, SP, LP and SLP. Once you are certain the drifting guidepost is set in its correct height, apply a small amount of bonding cement or glue to the base of each guidepost. Allow the glue to set before testing the performance of your adjustment. A clear playback picture should now be present in each of the various playback speeds.

Located on the lower side of the NEC VCR is a single drive belt which can be accessed by removing a metal belt guard. Also, the solder leads for the take-up sensor are now accessible as well as the lower side of each guidepost. It is necessary to access each guidepost from this angle in order to perform the above guidepost adjustment procedure.

9.3.A A Special Case:
The NEC Carriage Assembly Loading Gear
The VCR is really an ingenious invention. Each gear must be aligned correctly with its adjacent gear; the assembly drive mechanism must rotate at the proper play or record speed; and the carriage assembly is meticulously assembled to provide hours of glitch-free use.

VCR manufacturers have used nylon parts since the first VCR entered the United States. Often, a small nylon part is the culprit behind many symptoms in VCRs. Nylon parts are particularly vulnerable to the mechanical stress generated by the VCR. A nylon gear may crack, a transfer arm may come loose from its metal pin, or a small tab on the carriage assembly may sheer off due to excessive tension. Any one of these problems could make it difficult for the VCR to accept or eject videotapes.

As a rule, it is necessary to perform the following repair procedure if any one of the gears becomes defective. This procedure outlines the carriage assembly commonly found in NEC VCRs. This particular carriage assembly is tricky to service, but once you've mastered it, other carriage assemblies will seem easy to clean and repair. (*Figure 9-13.*)

Mounted on the right side of the NEC carriage assembly is a large spring-loaded gear. This particular gear is actually assembled from three individual gears that are held in place with a large tension spring. To access this large gear, several obstacles must be overcome.

First, remove the mounting screw that secures the loading switch housed on the side of the carriage assembly. This will allow the switch to be moved so you can extract and insert the new gear. Keep in mind, the nylon post on the large assembly gear should be located to the right of the loading switch when you remount the loading switch on the carriage assembly after the gear has been repaired.

Next, remove the nylon locking washer that secures the loading motor drive gear in place. This gear must be removed so you can extract and insert the new carriage assembly gear. Once the loading switch and loading gear are removed, the large gear assembly can be extracted from the side of the carriage assembly. The gear assembly is fastened to its mounting post with two nylon teeth. Gently pry these teeth open and lift the gear off its mounting post.

As you can see, this gear contains several areas which are susceptible to problems. If any of the following problems occur, they must be addressed and corrected if the VCR is to receive and eject a video cassette properly again:

1. The nylon post on the first gear may sheer off due to stress generated from its adjacent gears.
2. Several nylon tabs on the second gear may sheer off due to mechanical stress generated by the tension spring. (*Figure 9-14.*)

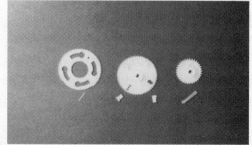

Figure 9-14. Broken tabs from damaged NEC gears.

3. The mounting post that secures the loading gear to the side of the carriage assembly may come loose and fall out while the VCR is in use.

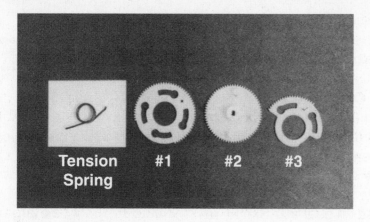

Figure 9-15. *Replacement gear parts.*

To properly reassemble this particular gear, it is a good idea to purchase the necessary parts beforehand and use the original faulty gear (while it is still assembled) as a guide. (*Figure 9-15*.)

Begin by placing the large tension spring on the center post of gear #2.   (*Figure 9-16*.) Notice the small hole located on gear #1. The end of the tension spring is bent at a right angle to fit into this hole.

It will be necessary to insert the straight end of the tension spring through the center hole of gear #1, to allow gear #1 to be seated on the posts of gear #2.

Figure 9-16. *Apply the tension spring onto gear #2.*

Once gear #1 has been applied to gear #2 and the tension spring is in place, simply rotate gear #1 so that it locks into place on gear #2. (*Figure 9-17*.) This has set the stage for gear #3. The straight end of the tension spring must also enter the center hole in gear #3 when performing the next assembly step. Once in place, gear #3 will be rotated and locked into place via the straight end of the tension spring. (*Figure 9-18*.)

Figure 9-17. *Gear #1 is on gear #2 with the tension spring in place, ready to receive gear #3.*

The final step requires gear #3 to be seated onto the entire assembly. Be certain to insert the straight end of the tension spring through its center hole. Once gear #3 has been seated on the nylon posts of gear #2, rotate gear #3. In order to retain this gear assembly, the straight end of the tension spring must be locked into place behind the nylon tab located on gear #3. Simply return the repaired loading gear assembly to its mounting post on the side of the carriage assembly by gently pressing down until its nylon teeth lock into place.

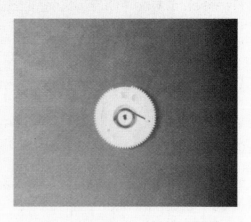

Figure 9-18. The tension spring is set in place on gear #3, locked behind a nylon tab.

Next, resecure the loading motor gear onto its mounting post with its nylon locking washer. If this mounting post is loose and slips in place, it's a good idea to use glue to keep it from falling out while the VCR is in use.

Finally, resecure the loading switch to the side of the carriage assembly with its mounting screw. Be certain the nylon post on the large gear assembly is located to the right of the contacts on this switch when returning it to the side of the carriage assembly. It is now possible to rotate the loading motor by hand to move the nylon post, if necessary.

Figure 9-19. The tape path is easy to access once the headshield is removed.

Reassemble the VCR and test the performance of the new carriage assembly gear. It should continue to receive and eject video cassettes for many hours of normal use.

9.4 RCA, Hitachi; Series Prefix VKT, VT

This particular VCR model tends to have a lot of potential hot spots which develop faults after many hours of normal use.

The overall layout of the VCR's tape path is relatively easy to access once a potential headshield is removed. (*Figure 9-19.*) Often, you do not have to remove the carriage assembly to perform an idler tire replacement procedure. However, when an End or Start sensor fails (mounted on either side of the carriage assembly), the carriage assembly must be removed and the faulty sensor will need to be replaced with a new one before the VCR will receive a video cassette properly.

In this particular VCR format, the symptom for a failed carriage assembly End or Start sensor might be that the carriage assembly cassette tray shifts forward then back again, about an inch, once the power to the VCR is turned on.

The idler drive assembly is often the pop-out type, and can easily be accessed through the top of the carriage assembly. Be very careful when accessing the idler assembly through the carriage assembly since the metal edges of the carriage assembly tend to be very sharp and unfinished. The tape counter belt can also be accessed through this area.

If the guideposts require adjusting, it is important to note that the adjustment nut located at the top of each guidepost requires a small Allen wrench to adjust. (*Figure 9-20.*) Be certain to always loosen the base locking nut of each guidepost first before performing an adjustment procedure.

Figure 9-20. *You should have Allen wrenches of various sizes on hand for adjusting guideposts.*

This particular VCR format incorporates six different drive belts, which are accessible from the lower side of the VCR once the lower pan has been removed. (*Figure 9-21.*)

A belt guard needs to be removed so you can access and replace the large flat drive belt as well as the large square belt. Located to the right of these two particular belts are two smaller loading belts. The smaller of

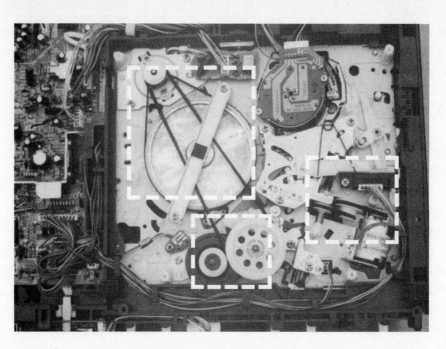

Figure 9-21. *Lower side drive belts.*

the two loading belts can be removed with a pair of tweezers. However, the loading motor mounting screw will need to be removed so you can access and replace the other loading belt.

When reinserting the loading motor, be certain the support tab located behind the motor on its mounting bracket is seated correctly in the chassis before tightening the mounting screw. (*Figure 9-22.*)

Keep in mind that there is a medium square belt located under the idler drive pulley that should be replaced as well. When this particular belt deteriorates and stretches, it tends to fall off and wrap around the other pulleys, thus prohibiting the VCR from running its Play or Record modes.

Figure 9-22. *Replacement loading motor, showing the location of the support tab.*

The belt kit for this VCR format is RCA part #199347. It is a good idea to keep this particular belt kit on hand since this format is very popular.

One particular area of this VCR where faults usually arise is in the power supply. Often there will be an STK voltage regulator IC secured to a heat sink. This particular IC regulates the different voltages used throughout

Figure 9-23. *A DC-to-DC converter.*

the VCR. It is considered to be a hot-running IC, which makes it susceptible to power surges and overheating.

The symptom for a failed STK IC might be continuously blinking power light. Also, the motors to the VCR may not initialize when the power supply cord is plugged into a wall outlet, and the upper head assembly may not rotate when the Play command is selected.

This VCR may also experience faults in the power supply fuses. Often there are as many as three or four different fuses incorporated within the VCR. A power supply fuse may fail after many hours of normal use or after a power surge. If this is the case, the fuse will need to be replaced with a new one before the VCR can work properly. There may be several white plastic caps concealing one or more power supply fuses. These may elude you at first, so be certain to inspect this location for a faulty power supply fuse.

Located toward the front display panel of the VCR, there may be a DC-to-DC converter. (*Figure 9-23*.) The DC-to-DC converter is considered to be a junction box for the VCR's clock display, and has a tendency to overheat and burn out. If this is the case, the clock display will no longer light and the VCR will be unable to record a preprogrammed TV show.

Figure 9-24. *Contents of a typical DC-to-DC converter replacement kit. (Courtesy of PRB.)*

The manufacturers have incorporated heat sinks into the newer DC-to-DC converters to dissipate damaging heat. Also, the DC-to-DC converter kits provide three capacitors, the DC-to-DC converter, and an IC protector. When replacing a faulty DC-to-DC converter, it is recommended to install every part of the new DC-to-DC converter kit to ensure that the fault will be entirely eliminated. (*Figure 9-24.*)

9.5 Phantom (Panasonic); Series PV-2400

WARNING: See section 9.6.C. **The Disposable VCR Challenge,** *for more information on servicing the Phantom Series PV-2400.*

Because this can be a complex VCR to service, some people consider it "disposable," though it doesn't have to be. However, you should order the service manual of this particular VCR since the realignment steps of the lower side cam gears are very complex. It is very important that the lower side cam gears are aligned in correct order for the VCR to work properly. Since the carriage assembly, mode/state switch and pinch roller mechanism are directly tied into the lower-side cam gears, the VCR will most likely default to the Stop mode if these particular gears are not aligned correctly. Given here are general servicing tips for this model; for more in-depth information, see section 9.6.C of this chapter.

The lid to this particular VCR format is often secured into place by two lower pan mounting screws. A locking foot system is used to keep the VCR lid secure to its plastic frame. Be certain not to break off these locking feet when removing the VCR's lid.

To access the various gears that may be out of alignment in the carriage assembly, it is necessary to remove the mounting screws that secure the main PC board in place. Once the mounting screws have been removed, the PC board can then be lifted to either access the video heads for cleaning or to remove the carriage assembly.

Located on the right side of the carriage assembly is a ribbon cable inserted into its connector. As a matter of design, the entire carriage assembly tends to "shake, rattle and roll" when a video cassette is inserted into the VCR. This continuous vibration tends to loosen the ribbon cable and the VCR may exhibit a symptom.

The symptom for a loose ribbon cable might be that the VCR suddenly stops during the Play mode or refuses to accept or eject a video cassette. Also, the VCR may accept a video cassette but does not go into Play, Review or Record mode. It is a good idea to hard solder this particular ribbon cable in place on the side of its connector to avoid this frustrating fault from occurring in the future.

To realign the lower gears of this particular VCR, it is necessary to remove the pinch roller, arm unit, sector gear, and pressure roller lift cam first. The entire pinch roller assembly is secured onto its mounting post by a removable nylon cap. It is a good idea to draw a simple diagram of the various parts in the order in which they are removed from the mounting post. Your diagram can then be used to ensure that the various parts are secured in their proper order once the lower side cam gears have been aligned.

Also, it is recommended that the pressure reduction and link gear that the carriage assembly sliding gear is seated over is also removed to ensure that it is correctly aligned with the lower side sub cam gear.

Once the pinch roller assembly, pressure reduction gear, and link gear have been removed from the top side of the chassis, it is often necessary to remove the entire array of lower side cam gears in order to realign each gear correctly.

Beginning with the sub cam gear, detent arm, and ring gear, proceed to align each gear with the various through-holes in the chassis and the alignment markings on each gear. Continue from the left side of the chassis to the right until all of the necessary gears have been inserted and are correctly aligned.

It is a good idea to secure the various gears into place with their mounting brackets when you are done, to ensure that the gears do not move when you insert the pressure reduction gear located on top of the chassis.

When the lower side cam gears are aligned and secured in place, it is necessary to insert and align the pressure reduction gear from the top side of the chassis. Align the pressure reduction gear with the through-hole on the sub cam gear. Once you are certain this particular gear is correctly installed, use your pinch roller assembly diagram and proceed to align the mode switch, pressure roller lift cam, and reduction gear with one another. It is important that the projection at the base of the lift cam aligns with the small hole on the pressure reduction gear. Also, the through-holes on both the mode/state switch and lift cam must also be aligned.

Next, the link gear and carriage assembly must be installed. It is important that the through-hole on the link gear aligns correctly with the through-hole on the sub cam gear. Simply secure the link gear in place on its mounting post with its nylon locking washer.

Depress the change lever and begin to rotate the center pulley counter clockwise. Observe the link gear on the top side of the chassis; it should begin to rotate. If the link gear does not rotate and the guideposts start to extend to their V-block position, you are rotating the center pulley in the wrong direction. While keeping the change lever depressed, proceed to slowly rotate the center pulley until the link gear stops rotating. It is a good idea to mark the tooth on the link gear that stops at the edge of the VCR's chassis. This tooth can then be used as a reference for the next procedure.

Once the link gear stops rotating and the reference tooth has been marked, rotate the center pulley in a clockwise direction until the link gear rotates one full revolution in the opposite direction and the reference tooth on the link gear is back to its original position. This is the position the link gear must be in when reinstalling the carriage assembly.

With the cassette tray of the carriage assembly in its full-eject position, simply insert the carriage assembly so that the second tooth of the carriage assembly gear rack seats onto the fifth tooth of the link gear.

Assuming that the above assembly steps were performed correctly, the gear rack on the carriage assembly should simply seat into place over the link gear. The cassette tray may need to be shifted slightly forward in order to be seated correctly onto the link gear.

Install the carriage assembly mounting screws and reconnect the carriage assembly ribbon cable. Reassemble the VCR and test its performance in all of its various modes.

9.6 "Disposable" VCR Models

Many VCRs rapidly hit (or so it seems) a last gasp for air after only a year or two of continuous use. Lets face it, VCRs have evolved into disposable machines. This subtle transition has taken us from the large "piano key" type VCRs that broke into homes across the country in the early 1980s, to the more streamlined, lightweight models of today. The cost, as well as the manufacturing quality, has dropped dramatically. Since 1990, VCRs have taken on dramatic mechanical design changes, and require much more patience in order to disassemble them and diagnose their problems. As with any unregulated market that lacks quality control, it has become commonplace for a VCR to break down well before the credit card used to purchase the unit has been paid in full.

With nylon parts replacing metal, the initial market price of $1400 for a VCR nearly fifteen years has been reduced to a "bluelight special" price, usually starting at $139 or below. Plus, the VCR business has been plagued with horror stories of $149 repair bills and business owners trying to justify otherwise simple repairs. Not to say that all VCR centers are out to make a quick buck, but who can you trust? Many VCR owners are now left to ask, "Is it worth it to have my VCR repaired?"

In many cases, it is; that is, if you repair the VCR yourself.

This section has been provided to help the novice VCR repair and cleaning person to quickly identify problems in today's inexpensive or disposable VCR models. Once you understand the common areas that develop symptoms, repairing your disposable model will be easy and less expensive than buying a new one.

9.6.A The Disposable Mitsubishi

The first model outlined in this section is the low-end Mitsubishi, which often exhibits a "tape eating" symptom. Looking to the upper right of the carriage assembly within the VCR, there are several transfer gears, a videotape pull-out arm, and the pinch roller. The pinch roller is mounted on a nylon pressure roller lift cam. The various gears mounted on the chassis, which make up the pinch roller assembly, have been aligned to manufacturer specifications. When a fault does occur in a mechanical format such as this, it often originates from the pull-out arm.

Manufacturers often use a graphite-based grease on the pull-out arm which tends to dry out and harden after only several years of normal use. When this occurs, the pull-out arm becomes sluggish and no longer returns to its spring-loaded full-stop position. (*Figure 9-25.*) Since the pull-out arm is responsible for delivering the passing videotape to the pinch roller and capstan post, the symptom might be:
1. The VCR is eating videotapes.
2. The VCR will not eject a videotape.
3. The pull-out arm is wrinkling the edge of the videotape .

Figure 9-25. A frozen videotape pull-out arm.

Once the graphite grease dries on the mounting post, the performance of the pull-out arm becomes inhibited. As a result, the nylon gears may go out of alignment, or a nylon tooth will chip off from mechanical stress. If this is the case, the pull-out arm will need to be removed for a good cleaning, and several gears may need to be replaced if they are damaged.

The pull-out arm can be removed from the VCR by simply removing its locking nut. Be certain to clean the center hole on the pull-out arm, as well as its mounting post, with nail polish remover. (*Figure 9-26.*) It is a good idea to apply a light coat of VCR grease or sewing machine oil to the mounting post. Also, when returning the videotape pull-out arm to the VCR, the height will need to be adjusted by rotating the arm locking nut.

It is a good idea to use a junker videotape when testing and adjusting the height of this arm. Once the height of the pull-out arm has been properly located, it should continue to return to its spring-loaded full-stop position.

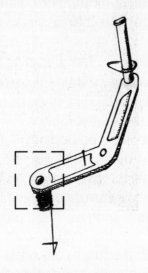

Figure 9-26. Drawing of a pull-out arm, showing the location of the mounting hole.

Once the pull-out arm and pinch roller have been removed, the next step is to be certain the relating nylon drive gears (mounted on the VCR chassis) are aligned correctly and have not lost any cogs or teeth due to mechanical stress. (*Figure 9-27.*) Manufacturers provide alignment holes on each gear to make aligning the various gears easier. Alignment holes or arrows are common to nearly every VCR on the market today. If any of the nylon gears is damaged, it must be replaced with a new one before the VCR will work properly.

Once you have made a visual inspection of this area, resecure the cleaned videotape pull-out arm to its mounting post on the chassis and return the pinch roller to the pressure roller lift cam within the VCR. Cross your fingers and test the performance of the videotape pull-out arm. It should quickly return to its spring-loaded full-stop position once a video cassette is either entered into or ejected from the VCR.

9.6.B The Disposable Fisher

In some inexpensive Fisher VCRs, the problem area tends to originate from the carriage assembly transport mechanism when a video cassette is entered into the VCR. Here again, it seems that the smallest nylon part in the VCR has been chosen to do most of the work.

Figure 9-27. *Nylon drive gears, correctly aligned.*

When a video cassette is entered into this particular VCR, a loading motor is initiated which rotates a large connecting cam gear.   In turn, this large cam gear rotates a smaller nylon gear which is aligned with a row of metal teeth. The row of metal teeth is a direct mechanical extension of the carriage assembly. As the motor and gears rotate, the metal row of teeth are extended forward thus receiving and lowering the video cassette into the VCR.   As you will see, the carriage assembly tray will need to be seated within the VCR, as if a video cassette had been entered, in order to perform this repair procedure.

Much like a hand saw, the row of metal teeth may work against the cogs on the small nylon gear if the drive belt on the connecting loading motor slips or fails. When this occurs, the symptom might be:
1. The small loading gear may lose a cog or tooth.
2. The VCR refuses to receive or eject a video cassette.
3. The VCR makes a loud grinding or "squelching" sound when a video cassette enters the carriage assembly.

To remedy this broken nylon gear problem, it will be necessary to disassemble the carriage assembly transport mechanism in order to extend the metal row of teeth forward, insert a new small gear, and realign the associated transfer cam gear.

Typically there are three distinct parts that make up the carriage assembly transport mechanism, which need to be removed in order to correctly perform this repair procedure. (*Figure 9-28.*) The three parts in question are:

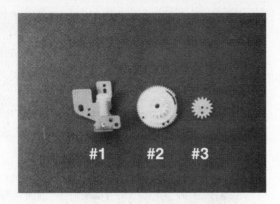

1. The loading motor worm gear bracket.
2. A large connecting transfer cam gear.
3. The small nylon link gear.

Figure 9-28. *Carriage assembly transport mechanism parts.*

Once you have successfully removed these transport parts, use a flat-head screwdriver to extend the metal row of teeth forward as far as they will go. (*Figure 9-29.*) Keep in mind the carriage assembly tray will automatically lower into the VCR as if a video cassette had been entered for playing or recording. Once the connecting nylon gears are correctly in place and the power to the VCR is restored, the cassette tray will automatically eject via the connecting gears, and your VCR will be as good as new.

After the cassette tray has been lowered and the row of metal teeth have been extended forward, simply mount the new small nylon gear onto its mounting post. (*Figure 9-30.*) It should lock into place. Be sure to use

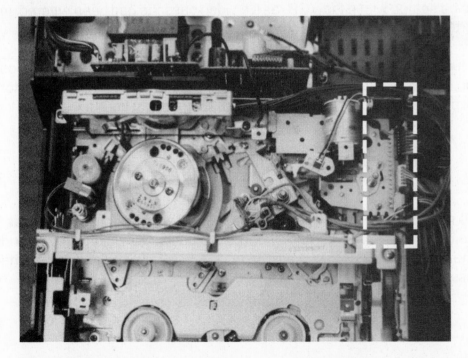

Figure 9-29. *The row of metal teeth is extended full-forward.*

Figure 9-30. *Insert the new nylon gear onto its mounting post.*

the alignment arrow cut into the gear to align the gear with the first tooth or cog on the metal row of teeth, before locking it into in place on its mounting post.

The next step requires the large transfer cam gear to be inserted then rotated until it locks into place against the small nylon gear. (*Figure 9-*

Figure 9-31. *The large transfer cam gear.*

31.) Begin by placing the toothless side of this large cam gear over the small nylon gear. It may be necessary to move the small PC board located next to this gear with light thumb pressure. This will allow the large cam gear more room to be seated onto its mounting post. When the large cam gear has cleared the PC board and is on its mounting post, slowly rotate the gear downward in a counterclockwise direction, until it drops and locks securely into place. It should now be possible to raise and lower the carriage assembly tray by simply rotating this gear by hand if it has been installed correctly. Be certain to rotate this gear so the row of metal teeth have returned to their full-forward position, as stated above, before continuing with the next step.

Finish this repair procedure by remounting the connecting loading motor worm gear bracket and dew sensor. The carriage assembly transport mechanism should now be repaired and ready for use. Plug the VCR in and power up. The cassette tray should eject on its own as the VCR initializes the various motors. Insert a video cassette into the VCR to ensure it receives and ejects well.

9.6.C The Disposable VCR Challenge

The Phantom PV-2400, the most complex of the disposable VCR models, is manufactured under several names, such as Magnavox, Sylvania, Philco and Panasonic. This mechanical format tends to have a mind of its own, and often exhibits symptoms without warning . If you are the type of person who likes mechanical jigsaw puzzles, this repair procedure is for you.

Figure 9-32. Removing the PC board will allow easy access to the carriage assembly gears.

WARNING: *It is highly recommended that you order a service manual for the Phantom PV-2400 VCR since the realignment steps of the lower side cam gears can be very confusing. It is integral that the lower side cam gears are aligned correctly for the VCR to work properly.*

Keep in mind, if you have success and understand how to perform this particular repair procedure, each VCR of this type will typically be the same to repair.   The method used here can be applied to every model exhibiting these parts and problems. Believe it or not, your goal is to disassemble the unit until all of the necessary gears, mounting brackets, and transfer arms have been removed. Notice that the alignment holes are aligned correctly for each adjacent gear. This is an important point to take into consideration throughout this repair procedure.

Symptom: *The lower side cam gears, carriage assembly and pinch roller mechanism tie directly into the top mode/state switch. The mode/ state switch is mounted on the chassis just behind the pinch roller assembly. If the lower side cam gears are out of alignment, the VCR will often exhibit a symptom. The symptom for a the carriage assembly or lower side cam gear that is out of alignment might be:*

1. *The VCR defaults to an all-stop mode without warning.*
2. *The VCR suddenly stops while playing a videotape.*
3. *The VCR refuses to either accept or eject a video cassette.*
4. *The VCR guideposts die in the out position when the Eject command is selected .*
5. *The VCR exhibits a rat-tat-tat grinding sound when a video cassette is inserted.*

To begin this repair procedure, remove the VCR's lid, lower pan, front panel, and carriage assembly. Set them aside in a safe place away from your work area.   It is a good idea to remove the mounting screws which secure the large PC board on the top of the VCR. This will allow the PC board to be lifted to allow more room to access the various carriage assembly gears that need to be removed and aligned during this procedure. (*Figure 9-32.*)

Remove the following parts from the VCR. These are the only parts to be removed from the top of the VCR (*Figure 9-33*):

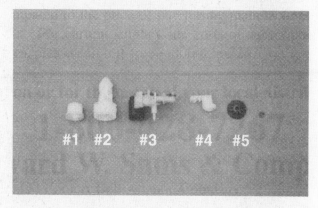

Figure 9-33. *Parts to be removed from the top of the VCR.*

1. Pressure reduction gear.
2. Pressure roller lift cam.
3. The pinch roller assembly.
4. Pinch roller locking cap.
5. The link gear with its nylon locking washer.

Figure 9-34. *Large drive belt.*

Next, turn the VCR over and remove the large drive belt which connects the capstan motor to the center pulley, which in turn rotates the lower side cam gears. (*Figure 9-34.*) To the left of the center pulley is the change lever. Depress the change lever and rotate the center pulley counter-clockwise. This will return the guideposts to their home position if the VCR has died in the out position, and will place the mechanical format into its full-eject position. This is where the guideposts and various parts must remain throughout this procedure if it is to be a success.

Figure 9-35. *Location of the span limiter bracket.*

Figure 9-36. *The sector gear unit (1), transfer arm (2) and sun gear (3).*

It is now time to disassemble the gears in order to realign them correctly. As you will see, nearly every gear is indirectly connected to the top mode/ state switch. When the various gears go out of alignment, so does the mode/state switch, thus creating one confused VCR.

Remove the span limiter bracket, which is secured in place with a mounting screw and two "C" locking washers. (*Figure 9-35.*) Begin removing the following from the VCR, in order:

1. Remove the sector gear unit, which connects the guidepost gears to their adjacent gears.
2. Remove the transfer arm, which is immediately to the left of the change lever.
3. Remove the sun gear, which is the large gear located between the sector gear and transfer arm. (*Figure 9-36.*)
4. Remove the main cam gear that the transfer arm runs within.
5. Remove the loading cam gear that the sector gear unit runs within.
6. Remove the retainer gear assembly, which houses three small nylon gears. (*Figure 9-37.*)
7. Remove the detent arm, which has a tab on either end to fit into the sub cam gear.
8. Remove the sub cam gear, which is notched to receive either end of the detent arm.
9. Remove the ring gear, which houses the retainer gear assembly. (*Figure 9-38.*)

*Figure 9-37. The main cam gear (4), loading
cam gear (5) and retainer gear assembly (6).*

If all has gone well to this point, the lower side of the chassis should be free and clear of all necessary gears. (*Figure 9-39.*) Now it is just a matter of reinserting each gear to ensure that their alignment holes are aligned correctly. (*Figure 9-40.*) This is the most crucial step to take into consideration when performing this repair procedure. It is a good idea to ensure each gear is not missing a cog or tooth before doing so. If a cog

Figure 9-38. The detent arm (7), sub cam gear (8) and ring gear (9).

Figure 9-39. *The lower side of the chassis, free and clear of all necessary gears, limiter arms and brackets.*

or tooth is missing from any of the gears, it must be replaced with a new one before the VCR will work properly again.   Be certain each hole correctly aligns with its adjacent gear when attempting to rebuild the lower side gear assembly.   VCR manufacturers use the method of marking each gear to ensure continuity.

Your goal is now to reassemble the lower side cam gears correctly so that you will be able to depress the change lever and rotate the center pulley. This will allow you to extend the guideposts to their full-forward position to ensure they are not inhibited in any way:

1. Insert the ring gear. Be certain its alignment holes match the alignment holes cut into the chassis.
2. Insert the sub cam gear so that its alignment hole matches the alignment hole on the ring gear.
3. Insert the detent arm so that its lower tab rests within the notch on the sub cam gear.
4. Insert the main cam gear so that its alignment hole matches that of the sub cam gear, now located underneath .
5. Be certain the retainer gear assembly through-holes are aligned with the holes on the ring gear, now located underneath.
6. Put the loading cam gear onto its mounting post. Notice it has an alignment hole that matches the alignment hole on the retainer gear assembly.
7. Insert the large sun gear over the retainer gear assembly and align the through-holes.

8. Insert the sector gear unit onto the loading cam gear. Be certain to align the through-hole located behind the first tooth at the tip of the sector gear unit, with the tab located on the loading arm T-unit connected to the right guidepost.

9. Put the transfer arm on its mounting post. The pin on this transfer arm runs within a groove cut into the main cam gear.

10. Return the span limiter bracket to its place over the sun gear and secure it with its mounting screw and "C" locking washers.

11. It is a good idea to secure the sector gear unit and transfer arm with their locking washers at this time.

12. Return the lower side drive belt to the center pulley and capstan motor.

13. Depress the change lever and rotate the center pulley in a clockwise then counterclockwise direction. The guideposts should mechanically extend to their full-forward position, then return to their full-eject position. The guideposts must be in their full-eject, stop, or home position before you install the top gear assembly.

The alignment hole on the pressure reduction gear aligns with the through-hole located on the upper left corner of the main cam gear. With the VCR still on its back, use a free hand to insert the pressure reduction gear and hold it in place while turning the VCR over so it is right-side up. This will ensure that the alignment markings are correctly aligned and the pressure reduction gear does not fall out when the VCR is turned to its upright position. Insert the pressure reduction gear at this time and turn the VCR over to install the remaining top side gears.

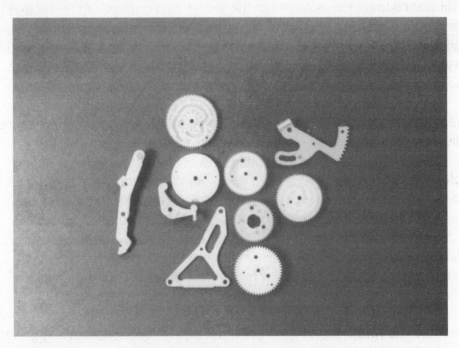

Figure 9-40. Each gear needs to be reinserted so that their through-holes are aligned properly.

With the pressure reduction gear correctly aligned and installed, the VCR should now be sitting right-side up on your work table. This procedure really comes down to how well each gear has been aligned up to this point. The top side gear assembly also has alignment markings which must be followed in order to achieve a successful result. (*Figure 9-41.*)

Begin by inserting the link gear so that its alignment hole is lined up with the hole on the sub cam gear. It is a good idea to secure the link gear on its mounting post with its nylon locking washer at this time.

Continue with the alignment procedure by maneuvering the transfer arm to insert the pressure roller lift cam so that its center notch correctly aligns with the markings on the pressure reduction gear. This is an integral point to take into consideration since the pressure roller lift cam is also aligned with the mode/state switch. Be certain the alignment hole on the mode/state switch is in line with the small hole on the pressure roller lift cam before installing the pressure roller lift cam. (*Figure 9-42.*) As a rule, this is the first alignment marking to check if this VCR begins to exhibit a symptom. If the holes on the mode/state switch and the pressure roller lift cam are not correctly aligned, the repair procedure to this point has been futile.

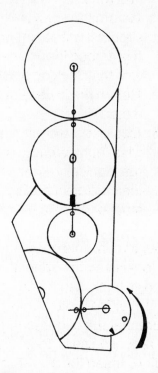

Figure 9-41. Drawing of a topside view of the link gear assembly. The arrow shows the direction the link gear needs to be rotated (counterclockwise).

It is a good idea to insert and secure the pinch roller onto the pressure roller lift cam with its nylon locking cap at this time.

Before inserting the carriage assembly, it will be necessary to depress the change lever and rotate the lower side center pulley counterclockwise until the unit is in its full-eject position. This is sometimes tricky to find. At any rate, the guideposts should be in their home position, and the alignment hole on the link gear should now be positioned to the right of the VCR chassis.

Once the mechanism has been set to its full-eject position, it is now necessary to rotate the link gear one full revolution; notice its alignment hole is now to the right of the chassis. To do this, it is necessary to depress the change lever while at the same time rotating the center pulley.

WARNING: Although the lower side cam gears may be correctly aligned and in position, it is a good idea to insert the pressure reduction gear at this time. The pressure reduction gear must be inserted from the top side of the VCR. This "missing link" connects all lower side gears with the VCRs top gear assembly.

WARNING: The link gear needs to be rotated one revolution in a counterclockwise direction before inserting the carriage assembly. It is a good idea to color one tooth on the gear with a felt marker where it comes in contact with the frame of the VCR, before rotating the gear. This will allow you to watch that particular tooth complete one full revolution to its original position, ensuring that the link gear has indeed rotated once.

Figure 9-42. *Be sure the mode/state switch and pressure roller lift can are aligned correctly before installing the pressure roller lift cam.*

It is a good idea to mark one tooth on the link gear with a felt marker before rotating it. This will allow you to observe the tooth make one full revolution. Once you have successfully rotated the link gear, the carriage assembly can now be entered into the VCR and secured into place.

Before seating and securing the carriage assembly into your VCR, the carriage assembly tray should be lowered as if the carriage assembly had a video cassette in it. This can be tricky at first, since you have to release several locks by hand that would otherwise have been released had a cassette been entered into the carriage assembly.

After the cassette tray is lowered, simply insert the carriage assembly into the VCR so that the second tooth on the rack gear is lowered onto the link gear. There is a rectangular through-hole on the carriage assembly that fits over the link gear. Be certain that this rectangular hole is seated flush on the link gear and that the teeth of the link gear mesh correctly with the rack gear on the carriage assembly. Secure the carriage assembly mounting screws into place.

Congratulations! Plug in the VCR and power it up. The carriage assembly should automatically eject, and the VCR should continue to work properly in all of its various modes.

Chapter Ten
HOOKUP METHODS
& SYSTEM LINKS

Chapter Ten
HOOKUP METHODS
& SYSTEM LINKS

Chapter Ten
HOOKUP METHODS & SYSTEM LINKS

This chapter is designed to walk the novice VCR owner through the many different antenna, cable box and television hookup methods with the least amount of confusion.

The main objective is to receive a television broadcast signal via an external antenna. This is a very important point to keep in mind since the diagrams for connecting the television, VCR and cable converter box become more and more complex with each new variation. Make a point to proceed slowly at each step, and try not to get confused or frustrated. If you think of your VCR, television and cable converter box in terms of simple line drawings with an "in" and "out" post on each component, the process of connecting the three into a working, cohesive unit will evolve more successfully.

Also, try to avoid randomly disconnecting and reconnecting the various coaxial cables midway through a particular hookup method hoping a clear picture will appear on the television screen. Guesswork does not often produce the results you're looking for. If you find yourself getting frustrated, stop! Take a deep breath and take a break from what you are doing. If you'd like, read ahead and try to decide precisely, from the diagrams in this chapter, which procedure is frustrating you. Once you have a pretty good idea of the different features each hookup method is capable of providing, don't hesitate to go for the one you want.

It is very helpful to begin the process of linking the various components together by starting with the type of antenna system you are using. Often, the television broadcast signal will enter your household via a rooftop antenna or through subscription television (cable TV) from either a backyard satellite dish or a curbside utility pole.

After you are certain of the antenna system you are using and the different types of connectors, proceed to link the various pieces of hardware together, one by one.

Before continuing any further, here is a list of the different connectors, tools and cables that are often used to connect a television, VCR and cable converter box so that you will receive a clear, recordable television broadcast signal through your VCR:

1. Matching transformer (300 ohm - 75 ohm).
2. Matching transformer (75 ohm - 300 ohm).
3. 5 ft. coaxial cable (75 ohm).
4. 5 ft. UHF twin-lead cable (300 ohm).
5. Coaxial cable into subscription cable box (75 ohm).
6. Coaxial cable signal splitter (75 ohm).
7. Coaxial A/B switch.
8. RCA video/audio connecting cable.
9. Rabbit ears antenna.

In order to outline the various types of hookup methods and system links, it is necessary to begin with a primitive antenna-to-television diagram. This diagram represents television before the advent of the VCR. The first antenna diagram was the mainstay for many years until the VCR, subscription cable, and satellite television were introduced. The viewing public now has many options to choose from in order to receive what was once a simple antenna-to-television broadcast signal.

As the different antenna, television and VCR models become more and more complex in design, each has its share of advantages and disadvantages. Keep in mind that the various hookup methods are designed to accommodate new technology in order to provide as many viewing options as possible.

However, with the advent of programmable remote controls, 125-channel cable-ready VCRs, video games, PIP (picture-in-picture), subscription cable, and satellite dishes, the overall theory and design of antenna-to-television has been augmented to the point where high technology seems to provide more confusion than viable viewing options.

10.1 Model #1:
Rooftop Antenna Signal to TV

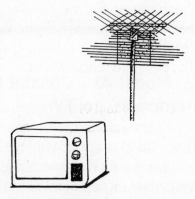

When the first VCR was introduced in the United States, it had to be incorporated into this original antenna scheme in order to receive the television broadcast signal through the rooftop antenna for recording purposes. (*Figure 10-1.*)

Figure 10-1. *Rooftop antenna signal to TV.*

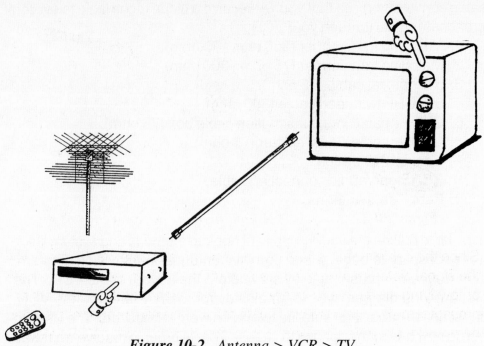

Figure 10-2. *Antenna > VCR > TV.*

10.2 Model #2: Antenna >VCR >TV

The external antenna must be connected to the "Antenna In" post on the back of you VCR so the VCR can receive broadcast television signals. The signal then exits the back of the VCR through the "Out To TV" post and enters the television set. (*Figure 10-2*.)

The next item to consider is cable television. Here a single coaxial cable is wired into the household from a curbside utility pole. Through this cable comes a coded transmission signal which requires a converter box to decipher and interpret.

With the advent of cable-ready televisions, rooftop antennas have all but been eliminated.

10.3 Model #3: Coaxial Cable from Street > "Antenna In" to TV

The coaxial cable enters your household from a curbside utility pole. This coaxial cable is then connected to the "Antenna In" post on the back of the television set. Keep in mind that in order for this scenario to work, you must subscribe to a cable carrier. (*Figure 10-3*.)

Figure 10-3. Coaxial cable from street > "Antenna In" to TV.

Since the cable signal is transmitted through a coaxial cable, rarely will the signal distort or fade. This hookup method is very popular in remote or low-lying areas where poor rooftop television broadcast signals are present. However, if you would like to record the cable programs provided by your local cable company, you need to patch your VCR into this configuration.

Keep in mind, in this particular hookup method, that the VCR may not be able to descramble some of the cable subscription channels.

10.4 Model #4: Coaxial Cable from Street > "Antenna In" on VCR > "Antenna In" on TV

The coaxial cable that carries the cable signal enters your household from a curbside utility pole. The cable then enters the VCR through the "Antenna In" post on the back of the VCR. The VCR can now be used to record various cable shows. The cable signal then exits the VCR through the "Out to TV" connector, and is then connected to your television set. (*Figure 10-4.*)

In order to truly descramble the incoming cable signal, a cable converter box must be linked to the system before the signal enters the VCR.

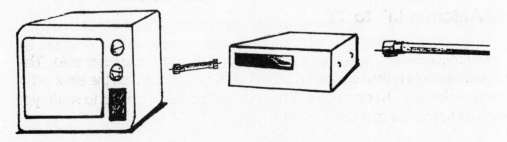

Figure 10-4. Coaxial cable from street > "Antenna In" on VCR > "Antenna In" on TV.

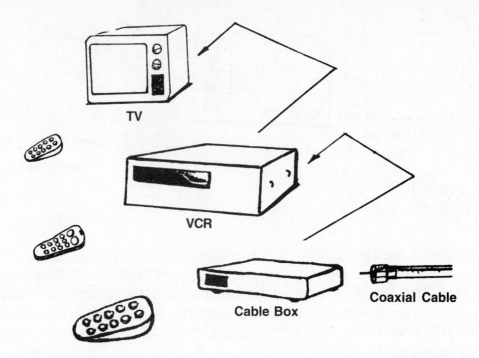

Figure 10-5. *Coaxial cable from street > cable box > VCR > TV.*

10.5 Model #5: Coaxial Cable from Street > Cable Box > VCR > TV

The coaxial cable enters your household from a curbside utility pole. The cable is then connected to the cable converter box, which descrambles the incoming signal. The descrambled signal then exits the converter box and enters the VCR through the "Antenna In" post on the back of the VCR. The incoming cable signal is now descrambled, and the VCR can now be used to record the deciphered signal. (*Figure 10-5.*)

One particular feature many cable subscribers enjoy is the ability to record a cable channel while watching another. This is usually not possible with the above hookup method; however, it IS possible if the cable converter box is placed AFTER the VCR.

As stated before, one drawback of this scenario is that the VCR may not descramble a specific incoming sport or movie channel that would have been descrambled otherwise if the coaxial cable had entered the cable converter box first.

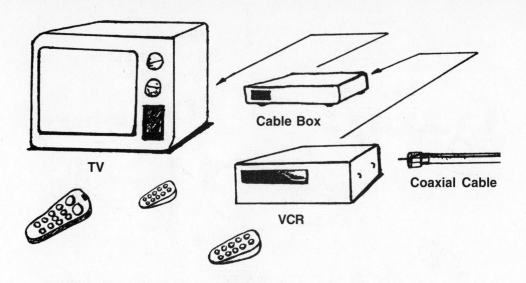

Figure 10-6. *Coaxial cable from street > VCR > cable box > TV.*

10.6 Model #6: Coaxial Cable from Street > VCR > Cable Box > TV

The coaxial cable enters your household from a curbside utility pole. The cable signal enters the VCR via the coaxial cable connected to the "Antenna In" post on the back of the VCR. Keep in mind that each specific channel must be tuned in through the VCR tuner. One drawback to this hookup method is that the VCR may refuse to receive or descramble a specific sport or movie channel which otherwise would have been deciphered by the cable converter box.

The cable signal then exits the VCR through the "Out to TV" post and enters the cable converter box. The cable converter box can now descramble the various subscription channels. The signal then exits the cable converter box and enters the television set. (*Figure 10-6.*)

The next three diagrams are considered to be advanced hookup methods. An advanced hookup method often incorporates a signal splitter and an A/B switch. These are used to either split or divert the same incoming cable or satellite coaxial signal. The split or diverted signal can then be used in another part of the same system.

10.7 Model #7: A/... > Cable or Satellite "In" > Cable Box > VCR > .../B

In this example, the incoming cable or satellite broadcast signal first enters a signal splitter and is split in two.

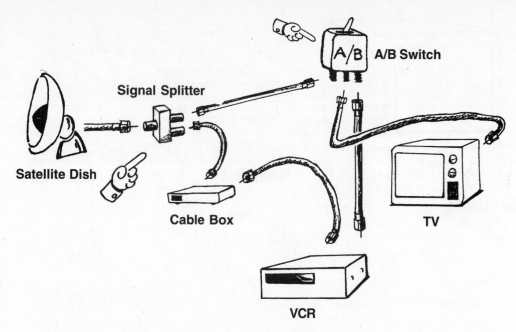

Figure 10-7. *A/... > cable or satellite "In" > cable box > VCR > .../B*

Upon exiting the signal splitter, one of the signals enters the A side to an A/B switch. The A/B switch is then connected to the television in order to view this signal. The signal exiting the other side of the splitter enters the cable converter box. This signal is descrambled and exits the cable converter box. The exiting signal enters the VCR through the "Antenna In" post on the VCR. The VCR is now capable of recording the descrambled cable signal.

The signal then exits the VCR through the "Out to TV" post and enters the B side of the A/B switch, which is connected to the TV. (*Figure 10-7*.)

This type of hookup allows you to record a descrambled program while watching another. Keep in mind that unless your television is cable-ready, it may not be possible to view a descrambled channel without going through the cable converter box on the B side of the A/B switch.

10.8 Model #8: How to Hook up a Video Game

Video Game... Cable "In" < > TV Cable Box > VCR > RCA Cables

In this example, it is possible to patch-in the necessary hardware for video games. In order to accomplish this, start by entering the incoming cable or satellite signal into a signal splitter. The incoming signal will now be split in two.

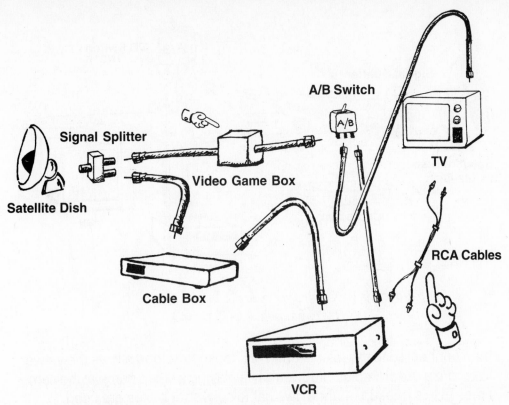

Figure 10-8. *Video game... cable "In" <> TV cable box >*
VCR > RCA cables.

One of the signals exiting the splitter connects to the video game box.
The cable exiting the video game box is connected to the television set.
The signal exiting the other side of the signal splitter enters the cable
converter box so it can descrambled. The descrambled signal exits the
cable converter box and enters the VCR through the "Antenna In" post.
The VCR can now be used to record the descrambled cable signal.
(*Figure 10-8.*)

The signal exiting the VCR can be connected to the television in several
different ways, depending on the type of connector on the back of your
television set.

The first method incorporates a set of RCA cables that connect the VCR's
audio and video output jacks to the television audio and video input jacks.
This allows both sound and picture to be transferred to the television from
the VCR.

The other method requires the video game box to be connected to the A
side of an A/B switch while the "Out to TV" post on the VCR is connected
to the B side of the same A/B switch. The signal exiting the A/B switch is

WARNING: *Some cable systems don't use boxes at all, but use add/delete filters for premium channels at the distribution boxes outside. In this case, assuming both the TV and VCR are cable-ready, the hookup is very simply "IN to VCR, OUT to TV."*

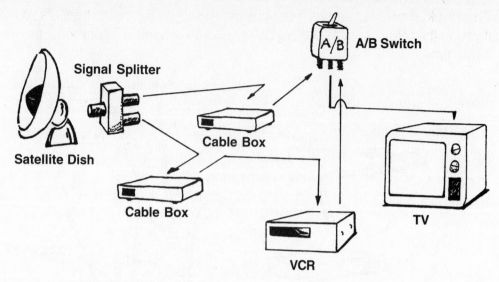

Figure 10-9. *Cable box > A/B... model cable "In" <>*
TV cable box > VCR > A/B.

then connected to the television. This hookup method allows the viewer to record a descrambled program while playing a video game at the same time. In this hookup method, two cable converter boxes are used.

10.9 Model #9: Cable Box > A/B... Model Cable "In" < > TV Cable Box > VCR > A/B

The incoming cable or satellite signal enters a signal splitter so that the signal is split in two. One of the signals exiting the signal splitter enters a cable converter box. The cable converter box can now descramble the cable signal.

The descrambled cable signal exits the cable converter box and enters the A side of an A/B switch. The A/B switch is then connected to the television set. The other signal exiting the signal splitter enters a cable converter box as well. The cable converter box can now descramble the cable signal.

The descrambled cable signal exits the cable converter box and enters the VCR through the "Antenna In" post. The VCR can now be used to record the descrambled cable program. (*Figure 10-9.*)

The signal then exits the VCR through the "Out to TV" post; but instead of going directly to the television set, it enters the B side of the A/B switch.

This hookup method allows a descrambled cable channel to be viewed through the television while the VCR is recording another channel at the same time.

Chapter Eleven
TWELVE ACTUAL
SERVICE CALLS

Chapter Eleven
TWELVE ACTUAL SERVICE CALLS

Now that you are familiar with the typical VCR format and have explored the workings of a VCR, it's time to consider starting a VCR repair service as added income and financial security.

Assuming that you are just starting out, chances are you do not have a lot of money for radio advertising, television commercials, or even a small storefront. However, it is possible for you to service VCRs on a part-time basis from the comfort of your home or apartment.

Keep in mind that the better you understand the information outlined in the previous chapters, the easier it will be to service and keep your customers.

The first important step after meeting your state's requirements for starting a business is to establish a "business" telephone line. In most cases, this will be required by your local telephone company if you intend to advertise in their telephone book. It is a good idea to speak with your local telephone book representative and obtain information for acquiring the business telephone line status. Often all that is required is your name, permanent address, and current telephone number. Also, ask the representative about the cutoff date for advertising in their telephone book. It will be a long, slow service year if you happen to miss the cutoff date by only a few weeks or even days.

Once you have established a proper telephone line, it is a good idea to invest in an answering machine for your business. (*Figure 11-1.*) This will allow you to receive calls while you are working or are across town doing errands. It is vital that your outgoing message on your answering machine contain the proper instructions for the first-time customer. It is very important to make the person who is calling feel that you are serious about repairing or at least cleaning their VCR. Try this outgoing message; it works!

> "Hello. Please leave your name and telephone number with a brief description of the problem with your VCR and I will return your call when I get in."

One line in particular that is very important to retain if you were to change this outgoing message is:

> "... a brief description of the problem with your VCR."

Figure 11-1. *An answering machine is a vital tool in a VCR repair business.*

Many times, when speaking of their ailing VCR, customers use only a handful of key words to describe the original symptom. It is a good idea to use this particular chapter to decipher some of the most common symptoms that calling customers will describe.

Here are twelve common symptoms as described by actual calling customers from my answering machine. Each description resulted in a service call. Most solutions to the problems outlined here are listed in Chapter Six unless otherwise indicated.

11.1 Actual Service Call #1

Symptom: VCR plays or records for only a few minutes before stopping.
Cause: Faulty idler drive tire.
Cure: Replace idler drive tire with a new one of correct size and shape.

Nearly every VCR on the market today contains a rubber idler drive tire located between the supply and take-up spool. (*Figure 11-2*.) The idler drive tire plays an integral role in the VCR's overall performance. After many hours of normal use, the idler drive tire tends to run smooth and begins to loose traction. When this occurs, the take-up or supply spool stops rotating at its proper speed, resulting in a fault.

Figure 11-2. *A replacement idler drive tire. (Courtesy of PRB.)*

Here the calling customer states, "The VCR seems to play for about five seconds, then it powers off."

What is actually happening here is that the idler drive tire has run smooth and is not making proper traction with the take-up spool. The customer has inserted a video cassette into the VCR expecting to view it. After depressing the Play command, the VCR loads the videotape through the tape path. However, since the idler drive tire is faulty and needs to be replaced, the take-up spool is not rotating at its proper speed. The VCR microprocessor notes an interruption in the pulse train being emitted from the take-up spool sensor, and commands an "all-stop" mode to the VCR motors as a safety feature to protect the videotape from potential damage.

In the second half of the same message, the customer states, "When I rewind the tape, I can hear it turning but it's not rewinding, and when I eject it, it eats the videotape."

What the customer hears is actually the worn idler drive tire slipping up against the supply spool in an attempt to rewind the videotape back into its plastic cassette shell. However, since the idler drive tire is worn and is no longer providing proper traction to either spool, the videotape is often left unwound in the VCR tape path. When the video cassette is ejected from the VCR, it is said that the VCR has eaten the videotape.

It is very important not to confuse the symptom of a slipping idler drive tire with that of a failed take-up spool sensor. If the take-up spool sensor fails, the VCR would not play for more than five seconds and the digital tape counter would not increment. Please refer to chapters Five and Six for information on accessing and correcting this problem.

11.2 Actual Service Call #2

Symptom: A loud "squelching" noise is heard from the working VCR. The VCR may load a video cassette but powers off shortly afterwards.
Cause: Faulty load/drive belt.
Cure: Replace all drive and loading belts.

Drive and loading belts are found in nearly every VCR on the market today. After many hours of normal use, they tend to stretch, become worn and eventually slip in place while attempting to execute a command selected from the VCR's front panel. (*Figure 11-3.*)

In this second example, the customer states, "Whenever we turn the VCR on to play a video, we hear a loud "squelching" noise coming from inside the VCR."

Figure 11-3. *Replacement drive and loading belts.*

The loud squelching noise is either a drive belt slipping on its pulley, or the belt is broken and the motor is just idling. When the owner turns the VCR on to play a videotape, the VCR attempts to initialize its various circuits by "setting everything to zero". When the signal reaches the drive or loading motor, the connecting drive belt begins to slip, which creates the loud squelching noise.

In the second half of the same message, the customer states, "...then the VCR shuts off. We can't use it the way it is."

Since the microprocessor notes that a particular motor is not initializing, and a loading belt is either slipping in place or has fallen off, the VCR powers off to the "all-stop" mode as a safety feature to protect the videotape from being damaged. If this is the case, the VCR will not work properly in all of its various Play or Record modes until the slipped belt has been replaced with a new one of correct size.

11.3 Actual Service Call #3

Symptom: There's no video; only audio when playing a videotape. "Snowy" picture, horizontal lines, black-and-white fuzz going across the playback screen.
Cause: Dirty video heads.
Cure: Clean the video heads thoroughly.

When a VCR begins to exhibit signs of dirty video heads, customers tend to explain the symptom by including how the VCR is performing, which has nothing to do with the dirty video head problem. For example, here the customer states, "The VCR fast-forwards and rewinds well..."

The mechanical functions of the VCR should be working fine since it is the individual video heads that have accumulated the dirt. In the second half of the message, the customer gives more insight to the actual problem: "We only get audio, and no video when we play any tape. There's just black-and-white fuzz on the screen." This is a good indication that the video heads have accumulated dirt.

Since the audio heads in a VCR tape path work independently of the video heads, the audio portion of the playing video tape is present and can be heard. Also, when there is "black-and-white fuzz" on the playback picture, this is an indication that the video heads have accumulated so much dirt that not even a portion of the playing videotape can be seen on the television screen.

To remedy this problem, be certain to give the video heads a thorough cleaning in order to restore the picture quality (*Figure 11-4*), as outlined in Chapter Four, before assuming the video heads are faulty and need to be replaced.

Figure 11-4. A replacement video head. (Courtesy of PRB.)

11.4 Actual Service Call #4

Symptom: The VCR immediately stops playing after the Play command has been selected. Also, the digital tape counter will not increment.
Cause: Faulty take-up sensor.
Cure: Replace take-up sensor with its exact replacement part.

Located beneath the take-up spool is the take-up sensor which electronically informs the VCR microprocessor whether or not the take-up spool is rotating at its proper Play or Record speed. (*Figure 11-5.*) When this particular sensor fails, the microprocessor notes an interruption in the information pulse train supplied by the take-up sensor. The VCR then shuts down as a safety feature to prevent further damage to either the videotape or the VCR. Here the customer states, "When we press the Play command to play a videotape, a 'still frame' comes on the screen."

Figure 11-5. A take-up sensor. (Courtesy of PRB.)

It is assumed that the owner has correctly entered the video cassette into the VCR and has pressed the Play command expecting to view the videotape. As the videotape is applied to the VCR's tape path, the upper head assembly which contains the video heads rotates at the proper speed. As the guideposts reach their fully extended position, the video heads begin reading the videotape and display the playback picture on the television screen.

However, the take-up sensor is faulty in this particular VCR. The VCR microprocessor notes this and sends out an "all-stop" command, which stops the take-up spool from rotating. Since the take-up spool is not rotating, the videotape is not being advanced through the VCR's tape path, which precedes the still frame on the television screen.

In the last section of the message, the customer states, "...then the VCR shuts down after we see the still frame appear."

Once the VCR microprocessor senses the faulty take-up sensor, the VCR is moved into an "all-stop" mode and often powers off to reduce the chances of damaging the playing videotape. The time it takes this entire process to occur is usually the time it takes the guideposts to reach their fully loaded V-block position.

With the VCR powered on, it is possible to rotate the take-up spool without removing the carriage assembly, in order to test whether or not the digital tape counter is incrementing. If the digital tape counter does not increment, chances are the take-up sensor is faulty. The faulty take-up sensor will need to be replaced with a new one before the VCR can work properly in all of its various Play and Record modes.

11.5 Actual Service Call #5

Symptom: Small "houses," or a "pac-man" ([]{[]}[]{}=[[]}]]) type maze, appear on the television screen when the VCR is on.
Cause: Faulty program IC.
Cure: Replace the program IC with a new one.

This faulty symptom does not need a lot of guesswork. The customer needs only to state several key aspects of the symptom. Nine out of ten times, the symptoms will point to a faulty program IC.

Let's look at the first segment of the caller's message: "I can't program my VCR to record a television show."

In this first segment of the incoming message, it is tempting to assume that the customer is having trouble with some aspect of programming their VCR. There are often many variables which the customer may be overlooking when attempting to program their VCR to record at a later time and date. However, let's consider the second half of the incoming message in order to draw a better picture: "We get what look like small houses or a 'pac-man' design on the television screen when we turn on the VCR on to use it."

This information tells us that the problem is beyond any misunderstanding the customer may have had when attempting to program their VCR. What has actually happened is that the program timer IC, which is responsible for interpreting and storing programmed information, has failed. This particular program IC is often located on the PC board toward the front of the VCR, behind the control panel. This program IC often fails in RCA brand VCRs. It should be simple to replace.

11.6 Actual Service Call #6

Symptom: The power light blinks for several seconds before powering off. The upper head assembly does not rotate when the Play command is selected. The VCR motors do not initialize when the power cord of the VCR is plugged into a power supply outlet.
Cause: Faulty STK voltage regulator IC.
Cure: Replace STK voltage regulator IC.

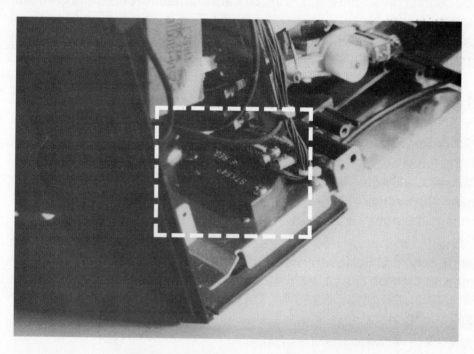

Figure 11-6. *Location of the STK IC.*

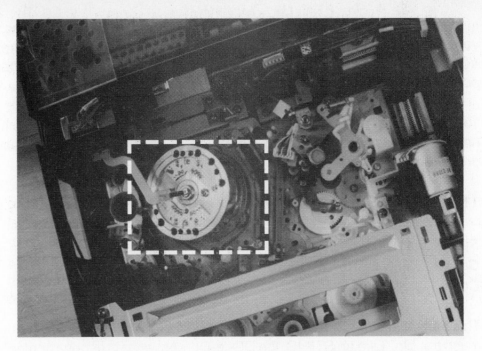

Figure 11-7. *The upper head assembly.*

In VCR formats which contain a STK voltage regulator IC, the IC is often located in or around the power supply. (*Figure 11-6.*) The STK IC is a voltage regulator and is responsible for regulating the various voltages that are used in the VCR. STK ICs are often very large, and are denoted by the letters "STK" followed by a specific part number. Always use the part number of the faulty STK IC when ordering a new one. Although STK ICs tend to look similar in size and shape, they are not interchangeable.

Let's look at the brief description of this symptom: "All of a sudden, the VCR stopped working."

Although we don't know that the problem originates from a faulty STK IC at this point, it is possible to deduct that the original problem may have originated from the VCR's power supply.

In the second half of the message, the customer states, "...so we took the lid off. Now when I play a videotape, the big disk in the middle doesn't turn, and the power light blinks on and off."

The "big disk" the customer is referring to is the upper head assembly. This is a common symptom when an STK voltage regulator IC fails. If an STK IC fails, the guideposts may extend to their full V-block position when the Play command is selected. However, the upper head assembly does not rotate, the videotape is not advancing, and the VCR soon powers off. (*Figure 11-7.*)

In this case, since the STK IC is a voltage regulator and is faulty, the correct voltage is not being applied to the upper head assembly circuits. The VCR microprocessor notes this and indicates a problem by flashing the power light on the front panel of the VCR for several seconds.

To remedy this problem, a new STK IC of the same part number will need to be installed. Refer to Chapter Six for this particular repair procedure.

11.7 Actual Service Call #7

Symptom: The front panel clock display is burnt out.
Cause: Faulty DC-to-DC converter.
Cure: Install a new DC-to-DC converter kit.

This particular symptom is in a class of its own, and is detectable without much trouble. Like an STK IC, the DC-to-DC converter is a "non-running" component which tends to fail without warning after many hours of normal use.

The DC-to-DC converter is often located just behind the front control panel inside the VCR. It is a vital link between the clock display and the other electrical components within the VCR.

The first segment of the customer's incoming call states, "The clock display on my VCR is burnt out. It doesn't count either."

When a DC-to-DC converter overheats, its inner circuits tend to burn out. When this happens, the clock display and digital counter often burn out as well. This is why the customer states that the VCR "doesn't count either."

In the second half of the incoming call, the customer states, "I need it to record my soaps while I'm not home." On some VCR models, if the DC-to-DC converter fails, it may still be possible to manually play and record televised programs. However, in most cases, the VCR cannot be programmed in order to record at a later time and date since the clock display is not functioning. This is the reason the calling customer is unable to record soap operas. In this case, a DC-to-DC converter kit will need to be installed before the VCR can work properly again. Chapter Seven describes the replacement procedure for this part.

11.8   Actual Service Call #8.

Symptom: The upper and lower edge of the videotape is being "cupped" or "scalloped". Lines appear either on the upper or lower portion of the playback picture.
Cause: The guideposts have come loose and are out of alignment.
Cure: Perform a guidepost adjustment.

Guideposts play an integral role in the VCR's overall performance. When a VCR is in the Play or Record mode, the guideposts are mechanically positioned forward to their full V-block position. The guideposts are used to extract the videotape from its plastic cassette shell and apply it to the various parts of the VCR's tape path.

Let's consider the first section to this customer's incoming message: "When we play any videotape, we see lines going across the screen."

Assuming this is not something a simple tracking adjustment could cure, through many hours of normal use, the guideposts may loosen and come out of alignment with the rest of the VCR's tape path. When this occurs, the symptom is often horizontal lines on either the upper or lower edge of the playback picture.

In the second half of the incoming message, the customer states, "Turning the tracking control knob doesn't help." When the guideposts begin to drift out of alignment, many times adjusting the tracking control knob will not correct the problem. This description is a good indication that the guideposts are at fault.

Another symptom which indicates that the guideposts are out of alignment with the other VCR tape path parts is tape scalloping, either on the upper or lower edge of the passing videotape.

In order to remedy these symptoms, a guidepost adjustment procedure must be performed, as outlined in Chapter Seven.

11.9   Actual Service Call #9

Symptom: Horizontal lines appear on the playback picture, and adjusting the tracking control knob doesn't help.
Cause: The audio control erase head is out of alignment.
Cure: Adjust the X-nut at the base of the audio control head.

The symptom for an audio control erase (ACE) head that is out of alignment is very similar to the symptom of several other VCR parts. (*Figure 11-8*.) If ever the term "symptom similarity" were to apply, it is here. The symptom of an audio control erase head assembly which is out of alignment with the other tape path parts resembles the symptoms for bad video heads, a misaligned guidepost, a faulty RF modulator, or a worn-out pinch roller. However, correcting the audio control erase head problem is much easier than correcting the others.

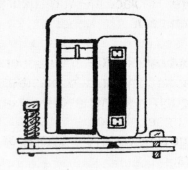

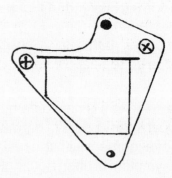

Figure 11-8. *Drawing of the top and side views of an ACE head.*

Let's consider the first segment of this customer's incoming message: "When I play any video, I have to turn the tracking all the way to the right." Here the customer indicates that when they attempt to play a videotape, the playback picture has some kind of snow or interference, so they try to correct it by turning the tracking control knob.

In the second half of the message they state, "I still don't get a really good picture; there are lines across the screen." From this new information, we can conclude that although the customer has tried to correct the original problem by turning the tracking control knob, horizontal lines are still present on the playback picture.

In this case, it is a good idea to first clean and test the video heads to be certain the problem is evident in each play speed, (SP, LP and SLP). Keep in mind that the video heads may be faulty. If the problem continues, perform an audio control erase head adjustment before assuming that the video heads are faulty.

11.10 Actual Service Call #10

Symptom: The VCR attempts to execute a front panel command but stops midway through, then powers off.

Cause: Broken worm gear, or the worm gear groove is obstructed.
Cure: Replace the worm gear with a new one or clean the groove on the worm gear.

Nearly every VCR on the market today uses a worm gear to complete specific commands selected from the front panel of the VCR. Sometimes, through many hours of normal use, the inner walls to the groove on the worm gear become soft and eventually collapse. Also, an object may lodge itself within the groove, or VCR grease may harden and accumulate here.

Often there will be a pin connected to a mechanical transfer arm that runs within the groove. If this pin is obstructed in any way, the VCR will be prohibited from completing the specified command selected from the front panel. If this is the case, a fault will occur, and the worm gear will need to be replaced with a new one.

Let's consider the first section of the incoming call: "We took the lid off our VCR, and it looks like the wraparound arms get stuck halfway after we press Eject."

Here the customer states that they have taken the lid off of their VCR and are looking for the problem. Maybe they have located something stuck within the VCR. However, when they state that the "wraparound arms" (they mean guideposts) get stuck halfway, this indicates that there is something more seriously wrong with the VCR than a simple object in the VCR's tape path, or they would have found it.

Let's consider the second half of this incoming call: "It doesn't let go, then the VCR powers off." In this last statement, the guideposts are holding onto the videotape since the worm gear, located on the underside of the VCR, is defective. In this particular case, the wall of the lower side worm gear had become soft and collapsed.

The mechanical transfer arm, which runs within the worm gear groove, is inhibited and cannot return the guideposts to the "home" position. Be certain the groove of the worm gear is cleaned, regreased, and free of any obstructions before assuming that the worm gear is faulty and in need of replacement.

11.11 Actual Service Call #11

Symptom: Televised programs through the VCR appear wavy and distorted in the playback mode.

Cause: Faulty radio frequency modulator (RF modulator).
Cure: Replace the RF modulator.

There are many variables to consider when an RF modulator fails. (*Figure 11-9*.) Always check the back of the VCR where the "In From Antenna" and "Out To TV" posts are located. If either post is loose, a fault may occur.

Figure 11-9. *A replacement RF modulator. (Courtesy of PRB.)*

Let's look at the incoming message for an RF modulator problem: "When we try to watch a television channel or a rental tape through the VCR..."

Here the customer states that they are either trying to watch a television channel or a videotape through their VCR. This is possible since the VCR has an RF modulator which is capable of receiving broadcast television stations through the "Antenna In" post on the back of the VCR. The VCR is also capable of playing a videotape, which exits the RF modulator through the "Out To TV" post.

In the second half of the incoming message, the symptom is described in detail: "...the screen is all wavy; not a clear playback picture at all." In this particular case, the reason the picture appears wavy is due to the fact that the "Out To TV" post on the RF modulator has become loose and is not transmitting a clear signal to the television.

11.12 Actual Service Call #12

Symptom: There is snow on the screen in one speed and not in another during the playback mode.
Cause: Faulty video heads.
Cure: Replace the video heads.

Faulty video heads are said to make up 11% of all VCR repair problems. Be certain to clean the video heads thoroughly before assuming that they are indeed faulty and must be replaced. Also, it is a good idea to use a video head tape speed tester to reduce guesswork. You can purchase one from your local parts supplier, or make one yourself using information from Chapter One, section 1.10, *Tricks of the Trade*.

Let's consider the first section of this incoming call: "The VCR plays our homemade videotapes clearly, but when we play a rental movie..."

Figure 11-10. Snow and lines on the TV screen as a
result of faulty SP heads.

Here we find that the customer is playing videotapes that they have made as well as rental movies on their VCR. Since rental movies are manufactured in a studio on a mass-market basis for the general public, the videotapes are often recorded in a playback speed common to most VCRs, usually the SP speed.

However, the typical home VCR is equipped with the option to record in several other speeds (SP, LP and SLP). This change in speed allows less or more recording space on a single video cassette, from two hours in SP to six to eight hours in SLP.

Chances are, in this particular case, the homemade videotape that the customer is playing was recorded in SLP in order to maximize the amount of time capable of being recorded onto the videotape. On the other hand, the rental movies were most likely recorded in the SP speed. Keep in mind that VCR video heads can be defective in one speed and not in another.

The customer goes on to state, "All I get is snow and lines when we play rented movies." The snow and lines are a result of the SP heads being faulty. (*Figure 11-10.*) The VCR may exhibit a clear picture in the LP and SLP speeds; however, until the video heads are replaced with new ones, any videotape that has been recorded in the SP speed will exhibit snow and lines.

Whenever replacing the video heads, be certain to use a video head extractor tool to ensure that the procedure does not incur another fault. Also, be certain to test the playback picture quality in each of the various record speeds. The replacement procedure is outlined in Chapter Seven.

Index

Index

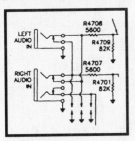

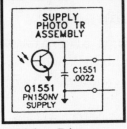
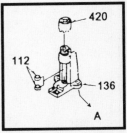

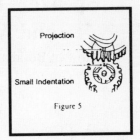
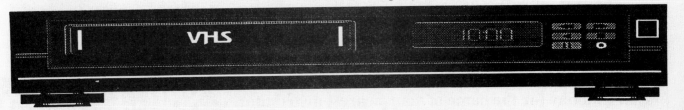